ALL YOU NEED TO KNOW ABOUT

ACUPUNCTURE

Also by Anthony Campbell

Medical Acupuncture: A Practical Guide
An illustrated companion to this book

So You Want to Try Acupuncture?
A book on acupuncture for patients

ALL YOU NEED TO KNOW ABOUT

ACUPUNCTURE

Anthony Campbell

Hawkmoth

First published 2013
Second edition 2016
Copyright © Anthony Campbell 2013, 2016
ISBN 978-1-326-86375-3
Published by Hawkmoth
www.acampbell.uk

Typeset using LyX on OpenBSD

Contents

Chapter 1

Introduction

Chapter Outline

▷ Symbols and abbreviations used

▷ Why I wrote this book

▷ Who is the book for?

▷ Modern medical acupuncture

▷ Felix Mann's acupuncture revolution

▷ Acupuncture in modern China

▷ How to use the book

1.1   Symbols and abbreviations used

 Danger—risk of injury

 Danger—risk of death!

Important departure from traditional acupuncture

Historical or academic interest (can be skipped)

Traditional acupuncture material

BMAS: British Medical Acupuncture Society
MPS: myofascial pain syndrome
TCM: traditional Chinese medicine
TENS: transcutaneous electrical nerve stimulation
TP: trigger point
VAS: visual analogue scale

1.2   Why I wrote this book

I have written two acupuncture textbooks previously so why another now? Several reasons. One is that it is fifteen years since the last book appeared. Quite a lot has happened in medical acupuncture since then and I wanted to bring things up to date.

Another is that both my previous books contained a chapter on traditional Chinese medicine (TCM), but by now the modern view of acupuncture is sufficiently well established, at least among British health professionals, that it no longer seems necessary to include a description of the ancient ideas. This is explained further in other sections in this chapter.

But the most important reason for writing now is that I want to focus even more strongly than before on the practical aspects of acupuncture. So I relegate discussion of the mechanisms of acupuncture and research evidence for acupuncture to chapters at the end of the book. This is not because I think these things are unimportant—quite the contrary—but because I expect that most readers will mainly be interested in how to do the treatments and that is what the book is intended to describe.

1.3   A change in format

It is not only in its content that the book is new but also in its arrangement. People read today differently from how they did in the past. Their reading has been shaped by the Internet, with a preference for shorter sections rather than longer explanatory paragraphs. So that is the model I am following here, which means that the book is more like an instruction manual than a textbook.

1.4   Who is the book for?

The book is written for people with a clinical background. I assume that you are a health professional (such as a doctor, physiotherapist, osteopath, chiropractor, nurse, or podiatrist) who is interested in acupuncture from a practical point of view. But I suppose it hardly needs saying that no book, including this one, can be a substitute for hands-on experience. This can only come from a practical training course. But there is a lot to take in on such a course and after completing it you need to think about what you have learnt and let it sink in. That is where a book like this comes in.

1.5   Based on teaching experience

In topic sequence and content the book is based on the acupuncture courses I have been holding for thirty-five years. It is designed primarily for students who have completed my course, but I hope it may also be useful if you have previously learnt acupuncture in another way. If that is the case you may find that the book offers a different way of thinking about what you are doing.

1.6   Avoiding dogmatism

This book describes my view and others hold different views. That is
as it should be. There are many ideas about acupuncture and no one
can or should claim to possess the Truth. My position is similar to that
of Bertrand Russell, who wrote: "I think we ought always to entertain
our opinions with some measure of doubt. I shouldn't wish people
dogmatically to believe any philosophy, not even mine."

I don't claim that what I describe is better than the other methods in
vogue, but it is at least as good and it is a lot easier and quicker to use
than most.

1.7   What you will *not* find here

Most of the exotic concepts you may have thought you would have
to acquire hardly appear at all in the text. For example, there will be
little to learn about 'acupuncture points' and nothing at all about the
so-called 'meridians '. (The Chinese term actually means 'channels'.)
This may come as a surprise, so I need to say something about the
modern (non-traditional) version of acupuncture.

1.8   Modern medical acupuncture

The kind of acupuncture I use and teach is based on the modern understanding of anatomy, physiology, and pathology. Historically, it
developed from TCM but now it has little connection with those ancient
ideas. How did this change come about?

1.8.1   A mainly British innovation

The new understanding of acupuncture arose in Britain in the second
half of the twentieth century, thanks largely to a British doctor, Felix
Mann (1.9). And Britain is still at the forefront of the modern approach
to acupuncture although similar ideas have been taken up elsewhere,
notably in Sweden. But while this way of understanding acupuncture is
new, in the sense that it makes use of the latest discoveries in physiology,

in a way it is a reversion to how acupuncture was understood in the West until fairly recently.

Acupuncture was used quite widely in Europe in the nineteenth century, but it had little if any connection with ancient Chinese ideas. Instead it was explained in the light of the medical theories of the day and it was sometimes not clearly distinguished from existing treatments like blood-letting and lancing abscesses. But there was a considerable change in how acupuncture was understood in France and Germany in the first half of the twentieth century, with a shift towards TCM.

1.8.2 A conversion to TCM outside Britain

The change in question was due to the efforts of a French writer, George Soulié de Morant (1878–1955). He had lived in China, where he became an enthusiast for TCM. On his return to France he translated a number of Chinese acupuncture texts and vigorously promoted the traditional ideas or at least his understanding of them. (This is where the term 'meridian' comes from.) So strong was his influence that today nearly all acupuncture in the West outside Britain is traditional even when it is done by scientifically trained health professionals.

1.8.3 Britain has a fortunate escape

This did not happen in Britain because by the twentieth century acupuncture was almost unknown here. It had flourished briefly in the 1820s but had then largely disappeared, so when Felix Mann wanted to study it as a young doctor in the late 1950s he had to go abroad to do so. The acupuncture he learnt was traditional because, as he said later, nothing else was then available.

He returned to Britain in 1959 and set up an acupuncture clinic in the West End of London—quite a bold undertaking at the time in view of the unfamiliarity of the treatment to British people. (He said he made contact with potential patients by taking people's pulses in the TCM way at parties!) He also began teaching acupuncture to doctors although few came forward to learn at first.

1.9 Felix Mann's acupuncture revolution

Initially Mann used traditional ideas, but he had a critical scientific mind and was prepared to experiment, putting needles in 'wrong' places to see what happened. He found that this worked just as well as putting them in the 'right' places and this discovery led him eventually to reformulate acupuncture in a modern way. He came to regard it as a means of modifying the activity of the nervous system.

Many of his departures from traditional practice are widely accepted today by modernists, even if they didn't learn their acupuncture from him directly. He can therefore be regarded as the father of modern medical acupuncture (but see 1.13 below).

1.10 A changing climate of opinion

By the 1970s opinions of unconventional treatment, especially acupuncture, were changing and more doctors wanted to learn it. Mann was the only doctor teaching acupuncture in Britain at the time so most of those who studied it did so with him; this was how I encountered it myself in 1977. Those who had studied under him constituted an informal medical acupuncture society. This group was the nucleus of what became, in the 1980s, the British Medical Acupuncture Society (BMAS), which now has over 2000 members. All are health professionals.

1.11 Different views of the role of TCM

At present most British health professionals who use acupuncture do so in the modern way although there are a few who adhere as far as possible to the traditional version. But I don't want to give the impression that there is a rigid separation into two groups, traditional and modern. Many modernists do hold to at least some of the traditional ideas, such as the existence of acupuncture points.

What we have, therefore, is a spectrum of opinion, ranging from those who seek to preserve as much as possible of traditional ideas to others like me who ignore the traditional ideas pretty well completely.

1.12 My reaction to Felix Mann's ideas

Felix's uncompromising rejection of TCM was not what I expected when I first looked into acupuncture in 1977. I did so because I was at the time interested in Oriental ideas and wanted to learn more about them.

Before starting Felix's course we were expected to read his books, which at that time were still based on the traditional system. They were pretty hard going and none of us, I think, made very much of them. But that didn't matter because the first thing Felix said to us was "I don't believe this stuff any more". And he went on to explain his reasons for this, which I have already alluded to.

Incidentally his rejection of tradition was not due to ignorance; he had gone deeply into the ancient roots of acupuncture and had even, with the help of Oriental scholars, read some of the classic Chinese texts.

I have to admit that my first reaction to his announcement was disappointment; it seemed something of an anticlimax. Later I was very grateful to him. I expect I should have come to a similar conclusion in the long run but he saved me a great deal of time.

1.13 Acupuncture in modern China

Since I wrote the first edition of this book a major study by Bridie Andrews has shed important light on the development of acupuncture in modern China. In her new book *The Making of Modern Chinese Medicine 1850–1960* (26.2.2) she shows that acupuncture as we have it today is quite recent; in fact it dates only from the 1930s! Its scope and its tools are now very different from what they were in earlier times, when it overlapped with minor surgery, used for draining abscesses and the like.

1.13.1 Cheng Dan'an

Acupuncture had fallen into disrepute in China among educated physicians in the nineteenth century. Its revival in the twentieth century was largely due to Dr Cheng Dan'an, who introduced ideas from Japan,

including the use of fine needles and modern anatomical concepts. He held that the acupuncture 'meridians' were a functional system that included the nerves, blood vessels and lymph nodes described by Western anatomists. Most remarkably of all, he taught that acupuncture works exclusively via the nervous system. He thus anticipated Mann's view of acupuncture, although it is likely that Mann arrived at his opinion independently, on the basis of his own clinical experience.

Acupuncture in China today is still based on the ideas of Cheng Dan'an and is therefore modern in character, making use of 'Western' conceptions of anatomy and disease. Enthusiasm for TCM is largely a Western phenomenon.

1.14   How I see TCM today

I think Felix Mann (and Cheng Dan'an) were right. We need to be radical. Traditional acupuncture is a fascinating subject to read about but it belongs to a pre-modern era. Attempts to preserve parts of it by reinterpreting them in the light of modern science, as some people seek to do, seem to me to be mistaken. Better to make a complete break.

1.15   An analogy with alchemy and astrology

Modern acupuncture has much the same relation to TCM as modern chemistry has to alchemy or modern astronomy has to astrology. You need not study alchemy to become a chemist and you need not study astrology to become an astronomer. The same is true of modern acupuncture: there is no need to begin with TCM.

No doubt the ancient Chinese physicians made many correct observations but the ways they interpreted them were inevitably framed in the light of the prevailing philosophy of the time. They were valid in their own terms but they are no longer useful today.

1.16   The critics have a case

Critics of acupuncture constantly quote research which finds that it makes no difference whether the needles are inserted at classic acupuncture points or randomly and they are largely right. Most attempts to show that there are specific effects from needling classic points have failed. The most recent example comes from the large-scale GERAC trials carried out in Germany (Chapter 25).

While it would be going too far to say that it makes no difference at all where the needles are inserted, needle location is a lot less important than many people believe and there is often more than one way to treat a given condition. *How* you needle is often more important than *where* you needle. That is why you will meet very few named points in this book.

1.17 But why not none at all?

Good question. You might say that logic and consistency would require me to avoid this terminology altogether, but there are disadvantages in doing so.

    ▷ One is that the names sometimes offer a convenient shorthand to describe where a needle has been inserted. I could make up my own names as a replacement, as Felix Mann did at a later stage in his teaching; but the new names were cumbersome and meant nothing to people who had not done his course or read his books so they were never widely taken up.

    ▷ Another problem with trying to avoid the traditional names is that even a non-traditionalist will often insert a needle at a traditional point although probably not with the precision that a traditionalist would expect. (I was recently asked to review a paper whose author defined the points he was using *to the nearest millimetre!*) Many acupuncturists, even those who regard themselves as modernists, use quite a number of traditional point names and they are likely to be confused if someone rigidly avoids doing so.

1.18 A working compromise

It may be that the traditional names will eventually fall out of use among modern acupuncturists but that has not happened yet. (Compare the continuing use of the ancient names of the constellations by modern astronomers even though they know that these bear no relation to the real structure of the night sky.) I shall therefore use a few traditional names from time to time—perhaps half a dozen in all. *This is simply for convenience, as shorthand. In no case does it imply any belief in their real existence as entities.*

1.19   Areas not points

Probably the most important innovation that Mann introduced was to abandon the need for great precision in needling. We should be thinking, he said, not of points but of *areas*, which may be quite large. (In my previous book I used 'acupuncture treatment area' to convey this idea.) If we speak of 'points' we imply, wrongly, the need for great precision in needling.

1.20   Naming the therapy

Another question is what we should call this form of treatment. It is often referred to as Western medical acupuncture and I think this can be justified on historical grounds, but here I shall use the term 'modern medical acupuncture'. Often this will be abbreviated to 'modern acupuncture' or just 'acupuncture'. Unless otherwise specified, in this book 'acupuncture' always refers to the modern non-traditional version.

1.20.1   Other names you may meet

Other terms you may have come across are 'intramuscular stimulation' and 'dry needling'. I take these to be simply different names for modern medical acupuncture. They were devised by people who wished to distinguish their use of needles from the traditional approach, but they tend to create confusion. I prefer to keep to 'acupuncture', which is, after all, a Western term—it is from the Latin and means simply 'sticking needles in people'.

1.21   No 'advanced' acupuncture

I am sometimes asked if I have courses in 'advanced' acupuncture. There are two answers to this, short and long. The short answer is that there is no 'advanced acupuncture'! Modern medical acupuncture is really an extension of your existing skills and there is no large body of esoteric knowledge that has to be acquired. The longer answer is that my approach is 'advanced' in at least two ways. First, you are expected to *think* about what you are doing and to adapt your treatment according to the nature of the underlying problem, so it does not depend on rote learning or rigid prescriptions. Second, it teaches periosteal (bone) acupuncture, which is a valuable technique that often does not figure in courses for newcomers.

1.22   How to use the book

This book has three main sections, describing (1) the basics of modern acupuncture, (2) the treatments I use, and (3) further reading.

1.22.1   The basics

Chapters 2–10 set out how I understand modern acupuncture. Read through these, since otherwise you will not understand the approach I use in the rest of the book. Chapter 3 is the most important of all; it is where I present what I take to be the essence of modern acupuncture. I believe this can be expressed in *four basic principles*. So far as I am concerned, everything that follows is an application and elaboration of these principles.

Safety is vital so the subject gets a chapter to itself (Chapter 6). Make sure you read this.

1.22.2   The treatments

Chapters 11–23 are about treatment. None of this is meant to be prescriptive and I expect that you will find other ways of treating various conditions. Indeed, I have learned a lot myself from people who have done my course: for example, the management of shin splints (20.3.5) and verrucas (21.8).

The arrangement of the treatment chapters is regional not systemic. That is, you will not find chapters describing how to treat musculoskeletal conditions, gastrointestinal conditions, and so on separately. Instead I take each anatomical area (head and neck, shoulder, upper limb etc.) and describe the symptoms and conditions that can be treated in each. For learning purposes I find that this works better than the alternative.

It is not essential to read these chapters in sequence. Not all of them will be relevant to everyone. If you are a podiatrist, for example, you need not read about the treatment of the head and neck. And some of the treatments, such as those for trigeminal neuralgia or ulcerative colitis, are useful for some practitioners but less so for others. Select those that are relevant to you.

1.22.3   Background reading

The final three chapters (24–26) are less directly connected with practice and could be omitted on a first reading. They are intended to give a wider perspective on acupuncture and include information about how acupuncture works as well as references for sections in earlier chapters and suggestions for further reading..

1.22.4   Marginal images

The images in the margin are meant to draw your attention to particular features in the text, especially potential dangers (skull symbols) but also ways in which the modern version of acupuncture departs significantly from TCM (bomb symbol). The owl symbol indicates that the section to which it refers is of mainly historical interest and is not essential reading if you are in a hurry. The yin–yang symbol marks a reference to TCM.

| If you read nothing else read Chapters 3 and 6. |

Chapter 2

Basic Acupuncture Facts

Chapter Outline

▷ **Acupuncture is a manual therapy**

▷ **What acupuncture is useful for**

▷ **Not a complete system of treatment**

▷ **Various kinds of acupuncture that you may encounter**

▷ **Treatments linked with acupuncture**

2.1 Introduction

Acupuncture can easily appear to be a strange exotic form of treatment, which makes it difficult to assimilate to the rest of clinical practice. It helps to dispel this impression if we note its similarity to other sorts of therapy with which we are more familiar.

Acupuncture has a lot in common with manual therapies such as osteopathy, chiropractic, and physiotherapy. The disorders they treat are quite similar and so are the kinds of effects that they produce. For example, osteopaths often say that their patients occasionally laugh or cry while being treated; the same is true of acupuncture. These similarities suggest that all the physical treatments may work in the same way—by modifying the activity of the nervous system. So we can picture all these treatments as overlapping with one another (Figure 2.1).

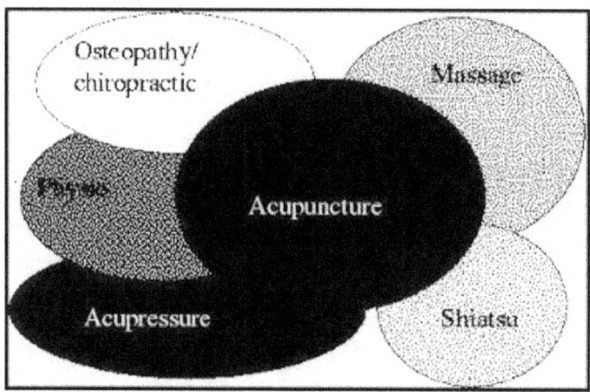

Figure 2.1

2.2 What is acupuncture used for?

2.2.1 Pain

The principal application of acupuncture is for the relief of pain. But it works better for some kinds of pain than others.

  ▷ Pain associated with tissue damage (good results)

  ▷ Musculoskeletal pain (good results)

  ▷ Neurogenic pain, e.g. trigeminal neuralgia (quite good results)

  ▷ Chronic pain syndromes, e.g. fibromyalgia (poor results}

  ▷ Central pain, e.g. post-stroke pain (very poor results)

2.2.2 Non-painful disorders

Acupuncture is sometimes used to treat non-painful disorders, such as allergies, nausea and vomiting, and some kinds of vertigo. How it works in these conditions is harder to explain than in the case of pain, but the assumption is that it depends on changes in the nervous system.

2.2.3 Addictions

Acupuncture has been used to treat various kinds of addiction, including to tobacco, and it has also been used to treat obesity (2.6.2). The evidence that it has a real effect in such cases is weak and I shall not discuss the treatment of addictions here.

2.2.4 Surgical analgesia

Following President Nixon's visit to China in 1972 the use of acupuncture to suppress pain during surgery attracted a lot of attention. In at least some cases it was apparently used instead of conventional anaesthesia. There is now doubt about the authenticity of this but it probably is true that acupuncture can inhibit pain in some situations. But this may not be an acupuncture effect in the ordinary sense. More likely it is similar to hypnotic analgesia, which can also be used instead of anaesthesia, though that is seldom done today.

2.3 Not a complete system of therapy

If you read some of the claims made for acupuncture by Western adherents of TCM you would get the impression that it can cure almost anything. This is far from the case. Acupuncture can be very successful in some circumstances but it is not a cure-all. In fact, it is not clear that acupuncture cures anything.

2.4 Not a cure

Acupuncture is best thought of a means of relieving symptoms, mainly pain. The evidence that it actually cures things, in the sense of reversing pathology, is largely lacking. If it is used to treat osteoarthritis, for example, it may relieve pain but is very unlikely to improve the damage to the joint or even to slow down its progress, at least as far as we know.

At the most, it seems possible that when acupuncture is used to treat a condition that may resolve on its own, the treatment may tip the patient into recovery mode or accelerate the process of recovery. But whether this really happens is difficult to prove.

2.5   How does it work?

There is a rational basis for modern acupuncture, based on what we know about the nervous system. In the 1970s the recently discovered endogenous opioids ('natural painkillers') appeared to provide a mechanism but nowadays they are thought to be only part of the explanation. Currently acupuncture is best seen as one of a number of treatments that modify the way the nervous system processes pain. It does so at several different levels.

 ▷ The local tissues

 ▷ The spinal cord

 ▷ The brain

Since I want to keep this part of the book focused on practical questions I don't consider the mechanisms of acupuncture further here, but there is more detail in Chapter 24.

2.6   Types of acupuncture

There are various types of acupuncture that you may encounter and also some other forms of treatment that are not acupuncture but are often associated with it.

2.6.1   Manual body acupuncture

The original form of acupuncture is what might be termed manual body acupuncture. This consists in inserting needles into the body at various places. This is the only form I shall describe in this book, mainly because I am unconvinced that the others add anything useful in practice. But you need to have at least a vague idea about the kinds that exist.

2.6.2   Ear acupuncture (auriculotherapy)

This is the best-known of several systems of acupuncture based on the idea that there are miniature representations of the body in various areas. It was developed in the 1950s by a French doctor, Paul Nogier. He believed that there is a map of the body in the external ear, lying upside down with the head at the bottom, the spine running up the border, and the feet at the top. He later invented various electrical machines for testing the ear and later still he described what he called the auricular-cardiac reflex: changes in the quality of the radial pulse when the ear was stimulated in various ways.

Practitioners who seek to treat addictions, including smoking, and to control appetite mostly use the ear.

I have experimented with ear acupuncture myself but, with one exception, I have not been impressed by the results. I am not convinced that ear acupuncture, or any of the other systems of the kind based on claims for miniature body maps, such as scalp acupuncture, offer any advantage over standard body acupuncture.

2.6.3   Electrical stimulation of the needles

This is an old idea. It was used by some French doctors in the nineteenth century and more recently it has been used by Chinese doctors to give prolonged stimulation for acupuncture analgesia in surgery. Some experienced acupuncturists find it valuable but my results have been less encouraging. Although I found good results when I used it for prolonged treatment in patients suffering from chronic pain, the results were equally good when I used my standard manual method with brief stimulation. I therefore gave up using electrical stimulation.

2.6.4   TENS

In this form of treatment electricity is applied to the skin via conducting pads rather than needles. It can give good pain relief to about a third of the patients in whom it is tried but the relief generally occurs only while the electricity is flowing or for a short time afterwards. Some patients who fail to respond to acupuncture may respond to TENS and vice versa. I have found TENS to be better than acupuncture to treat post-herpetic neuralgia.

2.6.5   Lasers

There is currently a vogue for using lasers instead of needles for acupuncture. This has the advantage that there is no risk of causing infection but there is uncertainty about the effectiveness of lasers. Even if they are effective, it is not clear that they work in the same way as acupuncture so it is somewhat misleading to refer to 'laser acupuncture'. It may be best considered as a form of treatment in its own right. I will not discuss lasers further in this book.

2.6.6   Electrodiagnosis

People who believe in the existence of classic acupuncture points sometimes attempt to find them electrically. The electrical resistance of the skin is supposed to be reduced at acupuncture points. The patient holds a neutral electrode and the operator tests various sites on the skin with an exploring electrode. The machine then sounds a buzzer or shows a visual signal of some kind when a 'point' is found.

The method is open to a number of objections.

2.6.6.1   Do points exist?

Obviously this method is only useful for people who think that acupuncture points are there to be found. Since I am sceptical about the existence of points I have no reason to look for them in this way.

2.6.6.2   Vulnerability to artifacts

It is easy to produce artifacts; if one keeps the probe in one place for longer or presses harder the deformation of the skin will alter its electrical resistance and hence the expected point will appear.

2.6.6.3   How deep are the points?

A further objection to the use of electrodiagnosis is that it depends on changes *in the skin*, but in the traditional system the points are supposed to lie at various depths *within the tissues*. It is not obvious why there should be differences in skin conductivity over these points—this is simply a convenient assumption.

2.6.7 Ryodoraku

This is a Japanese system of electroacupuncture which uses a special machine, the Neurometer. The electrical resistance is measured at sites on the wrist and ankle and then stimulation is applied to these places with the Neurometer. The method is not much used at present outside Japan.

2.7 Treatments linked with acupuncture

2.7.1 'Acupressure'

This is a rather unfortunate term, which ought to mean 'needle pressure'. In fact, it is widely used to refer to manually applied pressure to traditional points, usually with the fingers but sometimes with various kinds of apparatus. It has the advantage that the skin is not pierced so there is no risk of infection and little risk of damaging internal structures. It is often more painful than needle insertion and the effects usually last for a shorter time, but it can be useful on occasion; for example, if patients are very afraid of needles, or for self-treatment if you think self-acupuncture (Chapter 23) may be unsafe.

2.7.2 Cupping

This is an ancient form of treatment that was used not only in China but more widely in Europe and the Middle East. The method involves the production of a vacuum over the skin at the site to be treated. Traditionally this was done by lighting a piece of paper in a wine glass to heat the air; the glass was then placed on the skin, which was drawn into the glass as the air cooled and contracted. Nowadays, glass cups attached to a rubber bulb may be used instead.

Cupping is a counter-irritant and can relieve pain in that way, and there may be an additional effect due to the release of blood into the tissues to cause a local bruise. There is an old form of therapy known as autohaemotherapy, in which some of the patient's blood is withdrawn and then injected into the tissues. It is not clear how, or whether, this works, but if it does, cupping might have the same effect.

2.7.3 Moxibustion

This is a form of local heating, done in various ways, usually by burning plant material on or near the skin. It is a very ancient form of treatment; in fact, the earliest Chinese medical texts we know of, dating from about 200 BCE, describe moxibustion but make no mention of acupuncture. Incidentally, these texts describe 11 channels ('meridians'), not 12 as today, and they are not said to be connected to internal organs (26.2.3).

I have not used moxibustion myself and will not discuss it further here.

2.8   Conclusion

I have described several treatments that are associated with acupuncture, but although I have tried quite a number of them I always ended by returning to old-fashioned manual body acupuncture. I don't say that none of these other methods work but I am not convinced that they add anything useful. I regard acupuncture as a simple manual technique, and the less we complicate it the better.

Chapter 3

Principles of Acupuncture

Chapter Outline

▷ **All you need to know**

▷ **The four principles of acupuncture**

3.1 Introduction

Short though it is, this is the most important chapter in the book! It is where I explain what I take to be the essence of modern medical acupuncture. I arrived at this view in the first instance by analysing my own treatments. As I read other modern acupuncturists' descriptions of what they did I saw that they were using the same principles as I was, although without identifying them explicitly.

3.2   THE FOUR PRINCIPLES OF ACUPUNCTURE

> ▷ **Needle the site of pain**
>
> ▷ **Needle a remote site**
>
> ▷ **Needle the periosteum**
>
> ▷ **Needle for generalised (central) effect**

3.2.1   Needle the site of pain

This is the simplest kind of acupuncture and is often all you need. You insert one or more needles in the painful area itself. For example, widespread back pain, due perhaps to ankylosing spondylitis or osteoporosis, can be treated by needling the paraspinal muscles (16.2). Trigeminal neuralgia may be treated by needling the infratemporal fossa (13.7.3)

3.2.2   Needle a remote site

This depends on the phenomenon of referred pain. Trigger point acupuncture is one form of this. Another is the use of body segments (dermatomes, myotomes, sclerotomes) which is referred to as segmental acupuncture.

Examples of remote-site needling would be needling the ulnar side of the hand for upper thoracic pain (15.5) and needling an area at the back of the head to relieve a unilateral watering eye (13.4.6). Both of these have classic points associated with them (SI3, GB21).

3.2.3   Needle the periosteum

Needling the periosteum seems to have been introduced by Mann; it figures little or not at all in the traditional system. It is most commonly used to treat intrinsic joint pain such as that due to osteoarthritis, but can also act on wide areas of the body.

For example, periosteal needling of the articular pillar in the neck can influence symptoms in the upper half of the body (13.4.1), and needling the pelvic periosteum in the region of the sacroiliac joint can do the same for the lower half (18.6). See 10.14 for further details of the periosteal technique.

3.2.4   Needling for generalised (central) effect

In some cases acupuncture can produce widespread and quite profound generalised effects. This appears to be a central phenomenon, probably mediated at least in part by limbic system structures (24.4). Responses of this kind can occur no matter where the needles are inserted but the hands and feet seem to be most effective. I prefer to use the feet for this purpose—specifically a site on the dorsum of the foot between the first and second metatarsals (Chapter 12). This is known in the traditional system as Liver 3 (LR3).

> All the treatments I describe in later chapters are examples of how these principles work in practice, singly or in combination. In the words of John Keats, *that is all/ Ye know on earth, and all ye need to know.*

Chapter 4

Success in Acupuncture

Chapter Outline

▷ What you *don't* need to succeed

▷ What you *do* need to succeed

▷ What you need to acquire

4.1   Introduction

Becoming a good acupuncturist does *not* depend on learning large numbers of 'acupuncture points'. It is largely a matter of applying your existing professional knowledge in a different way.

4.2   The main requirements

▷ Anatomical knowledge

▷ Skillful palpation

▷ Alertness to patients' reactions

▷ Willingness to improvise

▷ Skillful needling

▷ Treating a lot of patients

The first three are skills that any health professional may be expected to possess already. The remainder need a few words of explanation.

4.2.1 Willingness to improvise

As students we learn various facts and theories, some of which will turn out to be wrong in the years that follow graduation. As we continue to practise we inevitably acquire our own ways of working, which will be—should be—different from what we learnt as students. This is true of acupuncture as well.

In the light of the four principles you will be able to devise your own treatments for conditions that I don't describe here. So you should never ask "Can I do x, y, or z?", because the answer will always be the same: *you can do it if it's safe.*

Done in this way, acupuncture becomes more creative and interesting—and more effective—than if you rely on what are referred to as acupuncture 'cookbooks'. These are really very simplified versions of TCM texts, with lists of 'points' to try in various conditions. Avoid them.

4.2.2 Needling skill

This does require a little practice. Inserting needles is not normally part of your repertoire if you are a physiotherapist, osteopath, or chiropractor, but your work depends on manual skills and it doesn't normally take long to learn to use the needles.

If you are a doctor, podiatrist, or nurse you will be familiar with inserting needles, but there are differences between the needles used for injections and those used for acupuncture. Acupuncture needles are solid not hollow and don't have bevelled cutting edges at their tips; they are also flexible. These things can make them more difficult to insert at first.

4.2.3   Treating a lot of patients

Like any other therapeutic technique, acupuncture needs practice to become good at it. The secret of learning acupuncture is to use it a lot in as wide a range of conditions as possible. You should consider the possibility of using acupuncture each time a patient walks in at the door.

> If you are a manual therapist your success rate with acupuncture will be about the same as with your existing treatments.

Chapter 5

Trigger Points and Acupuncture

Chapter Outline

▷ The historical background of trigger points

▷ The myofascial pain syndrome

▷ What are trigger points?

▷ Active and latent trigger points

▷ The status of the trigger point idea today

▷ Trigger points and acupuncture

▷ Locating trigger points

▷ Needling trigger points

▷ Fibromyalgia

5.1   Introduction

The TP story begins in the 1930s, when J H Kellgren, a rheumatologist, conducted important studies on referred pain in volunteers, including himself. He did this work when he was a young doctor at University College Hospital in London. The idea was suggested to him by his tutor, Sir Thomas Lewis, who had pioneered the notion of referred pain. Kellgren took this further. Similar work was being done independently at this time in Germany and Australia.

5.1.1   Inspired by acupuncture?

The connection with University College Hospital is interesting, because this was one of the last centres in Britain in which acupuncture was practised in the nineteenth century. Acupuncture had been used quite widely by British doctors in the 1820s but interest largely faded out later. The exceptions were the Leeds General Infirmary and University College Hospital, where it continued to be used until 1870. It seems possible that Lewis knew about this and that it prompted his interest in referred pain, although I don't know of any evidence to support this idea.

Kellgren investigated what was then called 'fibrositis'. Using hypertonic saline injections to irritate different anatomical structures, including fascia, tendon, and muscle, he demonstrated that pain was produced that had a characteristic quality and gave rise to specific referral patterns.

5.1.2   The myofascial pain syndrome

In the 1940s A. Steindler, an orthopaedic surgeon, coined the phrases 'myofascial pain' and 'trigger point'. These ideas were taken up a few years later by an American physician, Janet Travell, who spoke of the 'myofascial pain syndrome' (MPS). She was subsequently appointed as White House physician to both President Kennedy and President Johnson. She published, together with David Simons, a large two-volume work which is still the 'bible' of those who work in this field (26.2.5).

5.2   What are trigger points?

TPs are defined as areas, usually in muscles but also in ligaments and
musculotendinous junctions, that are tender when pressed and refer
pain and other sensations to distant areas (26.2.5).

To inactivate TPs Travell and Simons used injections of local anaesthetic
and also application of a cooling spray to the skin followed by stretching
the affected muscles ('stretch and spray').

They recognised both active and latent TPs.

  ▷ Active TPs give rise to symptoms.

  ▷ Latent TPs are tender when palpated but don't cause pain sponta-
    neously.

Active TPs may cause other effects in addition to pain. These include:

  ▷ Sensations of various kinds, including 'crawling' sensations in the
    skin

  ▷ Autonomic effects

  ▷ Muscular weakness

5.3   The status of the TP idea today

This work has given rise to quite a lot of research, but it has never taken
off in mainstream medicine and doctors learn nothing about TPs or
the MPS in their ordinary training. Physiotherapists, osteopaths, and
chiropractors do encounter these ideas, and this has led many of them
to study what is often called *dry needling*. The name refers to the fact that
a needle is inserted into the TP but nothing is injected (injection would
be *wet needling*). J take dry needling to be a synonym for acupuncture.

5.4   TPs and acupuncture

Although not everyone agrees, it is often said that almost all the TPs
in the Western literature have a classic acupuncture point nearby, and

about seventy per cent of the classic acupuncture points have a TP nearby. The traditional system recognises the existence of 'Ah Shi' ('Oh Yes') points, which are not part of the regular system of acupuncture points but which become tender in disease; these are probably the same as TPs.

5.4.1   How useful are TPs for acupuncture?

There is only limited evidence to show that needling TPs has a beneficial effect. A literature search in 2008 found that most published studies were of poor quality and only one supported this form of treatment. And a review of the criteria used by 'experts' to diagnose TPs found considerable variations. Nearly all the studies cited Travell and Simons as the authoritative source, but most failed to apply the diagnostic criteria used by these 'authorities' (26.3.1).

5.4.2   Have TPs been over-emphasised by acupuncturists?

The importance of TPs for modern acupuncture has been overstated by many people in the past, probably including me. I think they do exist and are quite often responsible for pain syndromes, but to equate modern acupuncture with needling TPs is going too far. Apart from anything else, needling areas that are not tender often works and such areas cannot be TPs because, whatever criteria one uses to identify TPs, tenderness is always one of them.

5.4.3   Are they 'points'?

I also have a problem with speaking of trigger *points*, which implies a greater degree of precision than is always needed. Trigger 'area' or 'zone' might often be more appropriate, and I shall use this terminology from time to time, but 'trigger point' is by now so well established that it would be perverse to avoid it.

5.5 Techniques for locating TPs

TPs are identified by palpation. This requires practice. Initially palpation should be gentle, since some patients are extremely sensitive to pressure, but in many cases firm deep pressure is required.

> ▷ **In flat muscles** there are often alternating taut bands and relaxed regions. A site of maximum tenderness will exist somewhere in the taut band; this is the TP. Sometimes, pressing it may elicit a muscle twitch. The palpation technique in flat muscles has been compared to feeling corduroy or plucking a guitar string.

> ▷ **In other, more rounded, muscles** you can find TPs by squeezing the muscle between finger and thumb. Again, there may be a twitch.

In both cases the patients will experience tenderness or pain when the TP is palpated. There may be radiation of sensations (not necessarily pain) from the TP on palpation. If this radiation corresponds to the areas in which the patient experiences pain, treatment is more likely to succeed.

5.6 Needling techniques

Once identified by palpation TPs can be needled. Note that the act of palpation can itself temporarily abolish a TP, so when you return with your needle the site can no longer be found. To avoid this you can draw a circle round the site of tenderness, but don't make a cross and needle through it, or you will tattoo the patient!

In practice I find it is possible to take a 'mental photograph' of the site in relation to skin blemishes or other landmarks, which avoids the need to mark the skin.

Visualise the depth and approximate size of the TP. Depending on the size of the muscle, I tend to picture something ranging in size between a pea and a walnut.

5.6.1   Decision time

At this point you have to answer two important questions.

▷ Do you needle directly into the TP (which may be at some depth) or do you just needle subcutaneously over it?

▷ If there are several TPs, do you needle all of them or just the most active ones?

There are no definite answers to these questions.

5.6.1.1   Needle the TP directly?

There are respected modern acupuncturists who always use superficial (subcutaneous) needling. Others do the opposite: they try to get into the TP and seek to elicit a twitch if possible.

5.6.1.2   Treat all the sites or just the most active ones?

Some sources say you should only do one or two sites, especially on the first occasion, while others maintain that you must needle every TP, otherwise it won't work. (This view tends to be held by those who favour superficial needling.)

5.6.1.3   My compromise view

My position on these questions is intermediate.

▷ I generally do try to needle the TP directly (provided it is safe, of course), but I don't try hard to get a twitch.

▷ I treat only one or occasionally two sites at the first session.

Over-treatment in either respect is likely to cause an aggravation. I use superficial needling in two circumstances:

▷ If the patient is apprehensive or is sensitive even to light manual pressure at the site.

▷ If there are vulnerable structures deep to the TP which I want to be sure to avoid.

5.7 Fibromyalgia

The MPS needs to be distinguished from fibromyalgia. The MPS produces fairly localised pain and there is often a history of over-use or injury to explain it. In fibromyalgia, in contrast, the pain and tenderness are widespread and have usually been present for a long time without an obvious cause.

5.7.1 Other differences

▷ Most fibromyalgia patients are women (90 per cent of cases).

▷ There are often associated symptoms affecting the bowel or bladder.

▷ Sleep is often disturbed, though it is unclear whether the pain affects sleep or vice versa.

▷ The cause of fibromyalgia is unknown but there does seem to be an association with psychological factors, although which is cause and which effect is unclear.

5.7.2 Response to acupuncture

The response in MPS is often good while in fibromyalgia acupuncture is much less successful. Any response that does occur is often brief and partial.

If you are treating a patient with fibromyalgia you should use particularly gentle stimulation because the patients are usually very sensitive to pressure and needles.

Chapter 6

Safety

Chapter Outline

6.1   Introduction

Acupuncture is generally safe, but it consists in the insertion of needles and that inevitably entails some risks. These should be small provided the acupuncturist is adequately trained and has a good knowledge of anatomy. But fatalities are not unknown.

To put this in perspective, the risks of acupuncture performed by a competent practitioner are probably smaller than those of taking a non-steroidal anti-inflammatory drug. These drugs are widely used and are available over the counter in many countries, including Britain.

The main risks of acupuncture are infection and anatomical damage.

6.2   INFECTION

Infection may be either viral or bacterial.

6.2.1   Viral infections

The two viral infections that are of particular concern in acupuncture are human immunodeficiency virus (HIV) and hepatitis B.

▷ **HIV** infection from acupuncture needles is exceedingly rare, although that is not a reason for complacency.

▷ **Hepatitis B** infection, on the other hand, has occurred many times. This is because the amount of blood needed to transmit infection is tiny.

Cross-infection of patients is prevented by the use of single-use needles. Acupuncture should always be performed with factory-sterilised needles that are disposed of after use. Provided this is done there is no risk of cross-infection from patient to patient via the needles. If the practitioner is a carrier of hepatitis B it is possible for the virus to be transmitted to the patient, though this is rare.

6.2.2   Bacterial infections

Bacterial infection can occur even though you use disposable needles. Bacteria may be present on your hands or on the patient's own skin. It is difficult to perform acupuncture while wearing gloves, so proper hand-washing or hand-wiping with alcoholic preparations is essential. Of the two, hand washing is probably more effective. Cleanliness is particularly important when diabetics are being treated because of the increased risk of infection.

6.2.3   Recognising infection

The classic signs of infection are:

  ▷ Local heat

  ▷ Local swelling

  ▷ Local redness

  ▷ Difficulty in moving or using the affected part

To these we can add:

  ▷ Red lines running up a limb, due to lymphangitis

  ▷ Tender swelling of lymph nodes in the affected region

  ▷ Fever

If any of these occur in the days after acupuncture you should strongly suspect infection and take steps to treat it.

6.2.4   Infective endocarditis

This is a rare but serious disease in which bacteria infect the heart valves and start to destroy them. The valves in question are usually abnormal in some way but you may not always know when this is the case.

Fortunately there is no risk of infective endocarditis with standard acupuncture. It may occur when needles are left in place for several days, as is sometimes done with studs in the ear. **Do not do this**. (Even the supposedly non-penetrating 'seeds' that are sometimes used as an alternative to needles may break the skin.)

6.2.5 To swab or not to swab?

Bacteria may be present not only on the patient's skin surface but below the surface in glands and hair follicles. The practice of wiping the skin with alcohol before inserting a needle is probably a meaningless ritual, and it has been abandoned by many hospitals and doctors. There is no need to do it for acupuncture unless the skin is visibly dirty. If you do choose to use a swab, leave enough time for the alcohol to evaporate or the needle insertion will sting.

6.3 ANATOMICAL DAMAGE

Probably every organ in the body has been damaged by an acupuncturist at some time. This is hardly surprising in view of some of the alarming treatments illustrated in some traditional books, which show long needles being inserted behind the sternum and in the orbit above the eyeball! Constantly keep anatomy in mind while doing acupuncture and err always on the side of caution.

The main structures to avoid are:

▷ Large blood vessels

▷ Large nerves

▷ The spinal cord

▷ Internal organs

▷ Joint spaces (may cause septic arthritis)

Remember that needling is never compulsory and alternative treatments always exist. The essential thing is to think in three dimensions: **is my needle point going into something that it shouldn't?**

I will review specific safety considerations for various areas in Chapters 11–23, where I take a regional approach to acupuncture, but two kinds of anatomical damage need to be discussed separately at this point: **pneumothorax** and **cardiac tamponade.**

6.4   Pneumothorax

This is probably the commonest serious injury due to acupuncture reported in the West. There is usually at least one example of it in Britain each year.

Needling anywhere over the chest wall is potentially dangerous; safe methods of doing this are discussed in 16.3.

6.4.1   Clinical features

The classic signs that a pneumothorax has occurred are:

▷ Pain

▷ Breathlessness

▷ Tachycardia

▷ Collapse

Note that these don't necessarily happen immediately; **they may take up to 24 hours to appear, and X rays are not always diagnostic immediately.**

6.4.2   Consequences of pneumothorax

Pneumothorax is not usually fatal but it may be: for example, if it is bilateral or if a tension pneumothorax occurs. In this, each time the patient breathes the amount of air in the affected pleural cavity increases owing to a valve-like effect at the site of the lesion. This produces a continual increase in pressure in the affected side of the chest. The heart and mediastinum are pushed towards the opposite side and the opposite lung is compressed. This is a surgical emergency requiring urgent treatment.

In a very emphysematous patient the functional loss of a lung owing to pneumothorax has occasionally resulted in death because the remaining lung was not able to sustain life.

6.4.3 What to do if a pneumothorax occurs

▷ Call an ambulance.

▷ Make a full record of the episode in writing; this will be essential evidence in case of subsequent litigation.

▷ Notify your defence organisation!

Since pneumothorax can take up to 24 hours to develop, if you think that there is even a small chance that you may have produced one you must warn the patient to go to hospital at once if symptoms occur in the next day or two. Make a written note that you have done this.

6.5 Cardiac tamponade

Some twenty years ago a patient in Scandinavia was having acupuncture at a classical TCM site over the lower sternum known as Conception Vessel 17 (CV17). She said, "I think I'm dying!", and indeed within two hours she was dead. Since then about a dozen similar cases have been reported in the West, fortunately not all fatal. The cause is cardiac tamponade.

6.5.1 Explanation

If fluid (e.g. blood) accumulates in the pericardium the heart ceases to fill in diastole and the patient dies. These patients had an ossification defect in their lower sternum which is found in 5 to 8 per cent of the population. It may be closed by fibrous tissue or a thin plate of bone but a needle can easily go through and reach the heart, where it may damage a vessel on the surface. This produces a leak of blood into the pericardium and cardiac tamponade ensues.

The sternal defect cannot be reliably detected by palpation or X ray and must be assumed to be there, so **the sternum should never be needled periosteally.**

6.6   Nerve damage

Although the location of major nerves is known, smaller nerves may be needled by accident. This is not completely avoidable.

Damage to nerves may cause pain, alterations in sensation, or (rarely) muscular weakness. In most cases nerve damage produces only transient ill effects, and patients should be reassured and told that the symptoms should disappear quite quickly, either the same day or a day or two later. **You should make a written note of such occurrences.**

6.6.1   Long-lasting effects?

Just occasionally, pain or other adverse effects due to nerve damage last longer than this—months or possibly even years. But this is very exceptional and could in principle occur after needle insertion for any reason, not just acupuncture.

6.7   OTHER UNWANTED EFFECTS

6.7.1   Needle phobia

Some people are terrified of needles, and these are obviously unsuitable for acupuncture. Not only are they liable to panic or faint, but acupuncture doesn't work in people who are very afraid of it. This has been noted in the Chinese literature over many centuries.

6.7.2   Fainting

Patients may faint when they are needled even if they are not particularly afraid of acupuncture. Those most likely to faint are young athletic men. If you think that a patient may faint do the treatment with the patient lying down.

6.7.3   Sweating

Some patients sweat a lot on their hands even if they are not particularly anxious. Give them a tissue to dry their hands if necessary.

6.7.4   Metal allergy

The needles, which are made of steel alloy, usually contain nickel, and if patients are allergic to nickel they may develop a rash round the site of insertion. This is not usually serious but nickel allergy is at least a relative contraindication to acupuncture.

6.7.5   Drowsiness

Some patients become quite drowsy after acupuncture, which is a danger if they are going to drive or operate other kinds of machinery. Patients should be warned about this and ideally should not drive themselves home after treatment, particularly on the first occasion. Sometime driving is unavoidable; in that case tell the patient to take extra care and make a note that you have done so.

As well as drowsiness, acupuncture can give rise to altered reactions. One man left the clinic in his car and promptly drove the wrong way round a roundabout, fortunately without an accident.

6.8   BLEEDING AFTER ACUPUNCTURE

Four types of bleeding may occur after acupuncture.

▷ **Capillary bleeding** is slight oozing of bright red blood. This is trivial to treat. Simply press the site with a tissue for a minute or two.

▷ **Venous bleeding** is similar to capillary except that the blood is dark. Again, all that is needed is pressure with a tissue for a minute or two. If the bleeding is from a varicose vein in the leg it may be profuse; in that case elevate the leg while applying pressure.

▷ **Arterial bleeding** does occur but is very rare. It may take the form of a jet of blood or, more likely, a tense swelling. Treat it by firm pressure kept up for about five minutes.

▷ **Haematoma:** Sometimes a small swelling arises when the needle is removed, without visible bleeding. It is about the size of a pea and is often felt rather than seen, especially if it is above the hair line. Press and massage the site for a couple of minutes to flatten it out. If you do nothing the swelling will take some days to be absorbed and may be uncomfortable.

6.8.1 Bleeding disorders and anticoagulants

Patients at particular risk of bleeding include those suffering from bleeding disorders, such as haemophilia or von Willebrand's disease, and those taking drugs that affect clotting (warfarin or the newer anticoagulants such as dabigatran). Aspirin and other anti-platelet drugs must also be considered.

6.8.1.1 Precautions in these cases

▷ Patients taking warfarin should bring a note of their most recent INR (International Normalised Ratio). This is a measure of their bleeding tendency. If the value is above 3 don't do acupuncture. Below this level acupuncture can be done but avoid deep needling, particularly in the limb compartments, for fear of causing a deep haematoma which could impede the circulation.

▷ In the case of the newer anticoagulants the INR will not be available. Acupuncture can be done with the same precautions as for patients taking warfarin.

▷ Many patients these days take low-dose aspirin bought over the counter in the hope of reducing the risk of heart attacks. Sometimes low-dose aspirin is prescribed by a doctor for this or a similar reason. Aspirin probably poses less of a risk than does warfarin. Even so, you should avoid deep needling, especially in the limb compartments, in such patients.

6.9   ACUPUNCTURE IN PREGNANCY

In earlier years there was a lot of concern about the safety of acupuncture in pregnancy, especially in the first three months, because it was thought to induce abortion. There was also talk about 'forbidden areas' in pregnancy. This advice was based on traditional sources.

6.9.1 Risk exaggerated or unreal

It now appears that the risks have been at least considerably exaggerated and may not exist at all. Nausea and vomiting in pregnancy have been widely treated without mishap (15.4). Large-scale trials of acupuncture to relieve back pain in pregnancy have been done in Scandinavia without causing problems (26.3.2).

6.9.2 Forbidden areas?

Chinese texts are not consistent in their account of the so-called forbidden areas in pregnancy, but one of these—the lower abdomen—should be avoided, and the same may apply to a site in the leg known as Spleen 6 (SP6), although this is more doubtful (20.4). SP6 has been used in clinical trials in pregnancy without ill effects.

6.9.3 Pregnancy summary

Looking at the question in a different way, we can say that if a patient requires treatment in pregnancy acupuncture may well be safer than using drugs, most of which carry some risk.

The decision to use acupuncture in pregnancy should follow discussion with the patient. Treat pregnant patients as strong reactors (7.3).

6.10 UNUSUAL ADVERSE EFFECTS

6.10.1 Broken needle

It is very rare for a good-quality needle to break. It is most likely to happen if a patient 'jumps' as the needle is inserted. To prevent this, always give a verbal warning first ("needle going in now ...").

If a needle should break, try to remove it with a pair of forceps or tweezers. If this fails the patient will have to go to their GP or hospital for removal. In that case, draw a circle round the site of insertion to make it easier to find.

6.10.2   Forgotten needle

It is surprisingly easy to forget a needle, particularly in women with long hair. Some of the needles have handles whose colour makes them difficult to see against the skin. The patient may not notice the retained needle until he or she is back at home. See 10.18 for ways to reduce the risk of this happening.

6.10.3   Nausea and vomiting

These are occasionally reported. You might try pressure or needling at the wrist if it happens in the clinic (15.4).

6.10.4   Laughter or tears

Occasionally, patients laugh or cry for several minutes when having acupuncture. The mechanism of this effect is unknown but it probably arises from events in the central areas of the brain that make up the limbic system. Similar effects sometimes occur with osteopathy or other forms of physical treatment.

Patients who react in this way should be reassured that it is something that is seen from time to time. I usually put it in a positive light, saying that it is a good effect which shows that the acupuncture has done something.

6.10.5   Seizures

Very rarely, patients have had epileptic-type seizures after acupuncture. (I have not seen this happen in nearly 40 years.) These are not usually major epileptic fits but take the form of loss of consciousness and some convulsive movements. Patients who produce such responses are not normally known sufferers from epilepsy.

The cause in most cases is a vaso-vagal faint resulting in temporary loss of blood supply to the brain. Episodes of this kind are actually quite common when people faint, although this is not much emphasised in the textbooks. It is also possible that seizures are sometimes due to stimulation of limbic structures by acupuncture, although this doesn't alter the favourable prognosis.

6.10.5.1   If a seizure happens

Reassure the patient, perhaps saying "You've had a funny turn" or something similar. Inform the patient's GP about the occurrence if you are not yourself the GP. With the patient's .permission (required) you should telephone the GP and give a description of the episode. Record it in your notes. Normally, no investigations are needed and the patient will not have to give up driving or take medication.

6.10.5.2   Try once more?

Should you ever try acupuncture again on that patient? In most cases the answer will be no, but there is no fixed rule. For example, if acupuncture was strongly indicated and if the patient had been seated on the first occasion, you might choose to do it once more with the patient lying down and using very gentle brief needling.

6.11   SPECIAL PRECAUTIONS

6.11.1   Oedema

Oedema may be of two kinds, pitting or non-pitting.

6.11.1.1   Pitting oedema

This occurs when there is excessive fluid in the tissues, caused, for example, by heart failure or kidney disease. Pressure with the fingers produces a 'dent', which slowly fills and disappears. Oedema of this kind is most evident in dependent areas, usually the feet.

Don't do acupuncture in such cases. If you do, there will be leakage of clear fluid which may go on for hours or days.

6.11.1.2   Non-pitting oedema

This is due to obstruction of the lymphatic drainage. It can affect the arms of women who have had radiotherapy for breast cancer, and it can also occur for unknown reasons ('idiopathically') in the leg.

Acupuncture has been said to have a risk of causing infection in non-pitting oedema, but this happens rarely and there is some evidence that it can relieve pain and even reduce the swelling in this condition. It therefore seems justifiable to use it for non-pitting oedema.

A mild degree of puffiness of the tissues after injury such as a sprain is not a contraindication to acupuncture.

6.11.2   Tumours

Avoid needling these.

6.11.3   Moles

Pigmented areas of skin might be malignant so should not be needled.

6.11.4   Areas affected by eczema or psoriasis

Avoid needling through these because they may be infected even if they don't appear to be.

6.11.5   Diabetic patients

Diabetic patients are at increased risk of infection so you should be particularly scrupulous with your hand-washing when treating them. These patients may have fragile skin, so avoid needling their legs if there is a risk of causing ulceration.

6.11.6   Patients who are overweight or underweight

Both these require extra care from an anatomical point of view.

▷ In fat patients the landmarks are difficult to identify.

▷ In thin patients there is only a short distance to go before you reach something important.

In both cases superficial acupuncture is often required. Most children require extra care as well, for the same reason.

To prevent anatomical damage, always think in three dimensions.

6.12 PUTTING THE RISKS IN PERSPECTIVE

Some of the risks mentioned in this chapter may seem a little alarming, but the available evidence for the safety of acupuncture from a number of studies is reassuring. For instance in 2001 the British Medical Journal published a survey by Adrian White and his colleagues of nearly 32,000 treatments given by doctors and physiotherapists which showed a low incidence of adverse effects, none serious (26.3.3). A survey conducted as part of the GERAC trials also confirms the low risk of serious mishaps (25.2.5).

6.13 CONTRAINDICATIONS: A SUMMARY

6.13.1 Relative contraindications

In some of these cases acupuncture may be possible with sensible precautions.

▷ Needle phobia

▷ Tendency to faint

▷ Immunocompromised patients (including diabetics)

▷ Skin infection at proposed needling site

▷ Eczema or psoriasis at proposed needling site

▷ Gross pitting oedema

▷ Patients about to drive or operate machinery

▷ Metal allergy

6.13.2   Absolute contraindication

Avoid electro-acupuncture for patients fitted with a demand pacemaker, for fear it may confuse the pacemaker.

6.14   Unsuitable patients

Certain categories of patients are unsuitale for acupuncture or require special precautions.

▷ Psychotic patients unless their symptoms are well controlled by medication

▷ Epileptic patients if their symptoms are not well controlled

▷ Needle-phobic patients

6.15   DANGER TO THE ACUPUNCTURIST

Infection of the acupuncturist from a needlestick injury is a constant risk, mainly for hepatitis B. **Acupuncturists are strongly advised to obtain immunisation against hepatitis B**. This greatly reduces the risk of infection although it does not guarantee immunity completely. Even if you have been immunised you should check your immune status periodically and always take precautions while needling to avoid injury (10.17).

If you do suffer a needlestick injury wash the wound with soap and water and cover it with a waterproof dressing before attending Accident and Emergency. Ask the patient if they are willing to attend also, for a blood test to establish whether they are a carrier of hepatitis B (but you cannot compel them to do so.)

Chapter 7

Phenomena of Acupuncture

Chapter Outline

▷ Patient's sensations

▷ Acupuncturist's sensations

▷ Strong reactors

▷ Pseudo-strong reactors

7.1 Introduction

In this chapter I describe what patients experience while having acupuncture and how you should evaluate these responses to help you adjust your treatment.

7.2 When needles are inserted

When an acupuncture needle is inserted several things may happen.

7.2.1   Initial pain

There may be a quick sensation of pain as the needle goes through the skin. This should pass off quickly; if it doesn't, withdraw the needle.

7.2.2   *De qi*

Next, the patient may feel various sensations which are difficult to describe because there are no words for them in English. They include feelings of heaviness, distension, or muscular stiffness that has been compared to the after-effect of exercise. They spread round the needle site to varying distances. They are referred to by the Chinese term *de qi*.

The term is also applied to sensations felt by the acupuncturist while manipulating the needle, such as resistance to twisting it. So this is a rather nebulous concept with a large element of subjectivity, but it is widely recognised by acupuncturists including r modernists.

7.2.3   Does *de qi* matter?

Qi is a term used in TCM to refer to a subtle 'fluid' which is supposed to flow in the channels; it has affinities with the Indian *prana* and Greek *pneuma*. There is no modern English equivalent although it is often (misleadingly) translated as 'energy'. In the traditional system acupuncture is supposed to alter the flow of *qi* in various ways and the *de qi* phenomenon indicates that this is has been achieved.

In the traditional system a lot of importance is attached to *de qi* and it is often said to be essential if the treatment is to succeed. But there is little good research evidence to show that this is true or even that obtaining *de qi* makes any difference to the outcome (26.3.4). Modern acupuncturists vary in how much importance they attach to it.

My opinion is that it certainly is not essential for the patient to experience *de qi* if the treatment is to work. Strong *de qi* may indicate that the patient is a good acupuncture subject although generalised (systemic) effects, discussed below, are probably more important.

7.2.4   Remote referral

When a needle is inserted anywhere the patient may feel sensations in a different part of the body, which may be remote from the site of insertion. For example a needle in the foot may produce feelings of warmth in the chest or throat. There may be accompanying flushing of the skin in the affected area, and I have seen a sharply demarcated red stripe appear in the neck after needling the foot. Such effects must presumably be evidence for the existence of unknown central pathways.

7.2.5   Generalised (systemic) sensations

Patients may report a variety of generalised sensations. These include relaxation, euphoria, and various alterations in perception. They are important because they may indicate that the patient is a *strong reactor* to acupuncture.

7.3   Strong reactors

The existence of a group of people who respond particularly strongly to acupuncture was first reported, at least in the West, by Felix Mann. It does not seem to be described in the Chinese literature, although one has to remember that probably only about a tenth of the Chinese texts have ever been translated into any Western language.

Strong reactors may experience a lot of generalised sensations, especially euphoria, and may even become quite ecstatic. If such people are over-treated, with too many needles left in for too long, they may have a bad reaction and feel quite ill for a day or two afterwards. But if they are treated very gently with minimal stimulation they may have excellent results. In a really strong reactor there may be a response in conditions for which acupuncture usually fails.

7.3.1   Recognising strong reactors

It would be useful if we could identify strong reactors in advance but this is not always possible. Mann said that they often have an artistic temperament or react badly to all the medicines that their GP prescribes.

I think they have a bright alert look. After one has seen several patients who turned out to be strong reactors one can sometimes pick them out in advance, though I should say with an accuracy of only about 70 per cent or less.

7.3.2   How common are strong reactors?

Estimates of the proportion of the population who respond strongly vary because it is not an all-or-nothing matter; rather we see a range of responses. Probably about 10 to 20 per cent could be classed as strong reactors, although a smaller proportion (1 to 5 per cent) would be ultra-strong reactors.

7.3.3   Pseudo-strong reactors

I recognise a group of patients whom I describe as pseudo-strong re-actors. They have a strong response the first time they are treated but on subsequent occasions little or nothing happens, much to the the disappointment of the patient (and the acupuncturist). These are pseudo-strong reactors; a genuine strong reactor will have more or less the same effect at each treatment.

Oddly enough, pseudo-strong reactors are often great believers in acupuncture when they first arrive. One would think that this is a good sign but in my experience it is not; these patients are often disappointed.

> Provided patients are willing to have the treatment there is no point in trying to persuade them that it will work. Disbelief is not a barrier to success.

7.4   Non-responders

Practically all the clinical trials that have been done in the West have found that a certain proportion of the population doesn't respond to acupuncture. There are different estimates of how large this proportion is but probably it is about one in five. This means that you are unlikely to get a better than 80 per cent response rate to acupuncture even in the most favourable circumstances. You should keep this in mind when telling patients about their chances of success with this form of treatment.

Chapter 8

Acupuncture Dosage

Chapter Outline

▷ **Needles are not drugs**

▷ **An acupuncture paradox**

▷ **What acupuncture 'dosage' depends on**

▷ **How to vary the strength of treatment**

▷ **Why brief needling works**

8.1   Introduction

It may seem odd to use the concept of dosage in relation to acupuncture since needles are obviously not drugs. But it is helpful to think about the amount of effect you are having by analogy to drugs.

Modern acupuncture can be thought of as a means of modifying the activity of the nervous system. So what matters is the number and type of nerve fibres that are affected. Skin, subcutaneous tissue, muscles, and periosteum all have different innvervations and produce different effects when needled but in all these tissues the concept of dosage applies. The following factors influence the 'strength' of the treatment.

▷ Thickness of needles

▷ Number of needles

▷ Amount of stimulation

▷ Depth of needling

▷ Duration of needling (if stimulation is used)

It follows that to get a bigger effect you can use thicker needles, insert more needles, give more manual or electrical stimulation, or needle more deeply. Conversely, of course, to get more gentle stimulation (usually desirable), do the opposite,

8.2 An acupuncture paradox

At this point we encounter a strange fact about acupuncture. Generally speaking, doing more does *not* equate to a better therapeutic effect. This is the opposite of what happens with drugs. For most drugs, to get a bigger therapeutic response we increase the dose although this increases the risk of unwanted effects.

For acupuncture things are different. Giving stronger treatment seldom improves the results and it can even give a worse response. To get a bigger effect, repeat the treatment at intervals but don't use stronger treatment. Sometimes, in fact, doing more has the reverse effect; some people seem to respond to gentle treatment but not to stronger treatment, odd though it may appear.

8.3 Few needles, brief stimulation!

In view of this paradoxrical response I generally use few needles (seldom more than four, possibly only one) and leave them in place for short periods only: a maximum of 1 to 2 minutes per needle, and often less—even 1 second in some cases! This is less than most acupuncturists  do. Traditionalists have got into the habit of using large numbers of needles and leaving them in place for 20 minutes or longer, and many modernists leave needles in for longer than I do. So why does brief work?

8.4   Why does brief needling work?

Well, why *shouldn't* it work? It is surely surprising that sticking needles into people should relieve their pain—I continue to be surprised by this myself, after nearly 40 years years—but given that it does, I see no reason why more prolonged needling should be more effective than brief stimulation. Anyhow, here are some reasons why brief needling works.

8.4.1   News of a difference

The nervous system notices changes in a stimulus but not continuity of stimulation. It registers *news of a difference*. For example we hear a sound when it starts but if its intensity and tone remain constant we cease to notice it after a few minutes. What we notice instead is its cessation. The same is true of acupuncture: the needle produces an effect when it is first inserted, but unless it is repeatedly stimulated either manually or electrically it soon ceases to have any effect. Most of the effect is produced in the first few moments after insertion.

8.4.2   Acupuncture is like throwing a switch

Much of the response to acupuncture probably depends on alteration of patterns of activity in the spinal cord and brain. This idea is discussed in Chapter 24, but for the moment we can think of acupuncture as a mechanism for switching pain processing off in the central nervous system. *Switching* is the important word here. When you turn on a light you don't have to keep your finger on the switch for twenty minutes. Switches are an all-or-nothing mechanism and the same is true of acupuncture.

Chapter 9

Managing the Treatment

Chapter Outline

- ▷ **Patterns of response**

- ▷ **Repetition of treatment**

- ▷ **Treating acute conditions**

- ▷ **When to give up**

- ▷ **What to do when response is brief**

- ▷ **Dealing with aggravations**

9.1   Introduction

Only three things can happen when patients are treated: they get better, get worse, or stay the same. I shall ll look at these in turn, starting with the most optimistic.

9.2 Improvement

This may occur as soon as a needle is inserted although that is rare. Even when it does occur the effect is sometimes brief; the symptoms may come back within half an hour or so. I think that if improvements take longer to come on they may last longer.

9.2.1 Delayed improvement

Improvement may appear later the same day or two or three days later. Some patients say that nothing happened for ten days but then all the symptoms disappeared. They are often kind enough to attribute this to your treatment but it is likely to be a spontaneous remission.

9.2.2 Partial improvement

The ultimate degree of improvement, usually after several sessions, may be complete remission or partial remission; but even a moderate improvement is worth having for many people.

9.2.3 Change in the tolerability of pain

Sometimes there is no great change in the amount of pain but patients find that they are less distressed by it than before. This constitutes a worthwhile improvement. The mechanism of this effect seems to be a reduction in activity in the cingulate cortex, which is the part of the brain that registers the emotional response to pain. (See Chapter 24 for further details.)

9.3 Build-up effect

A characteristic feature of the response to acupuncture is that the degree of improvement increases as the treatments are repeated. Most patients require more than one treatment to get the maximum benefit. So a course of treatment may consist of somewhere between two and six sessions, though a few patients may be completely cured by one treatment and others may require more than six.

9.4   Diagrammatic representation

Figure 9.1 illustrates the possible response patterns diagrammatically.

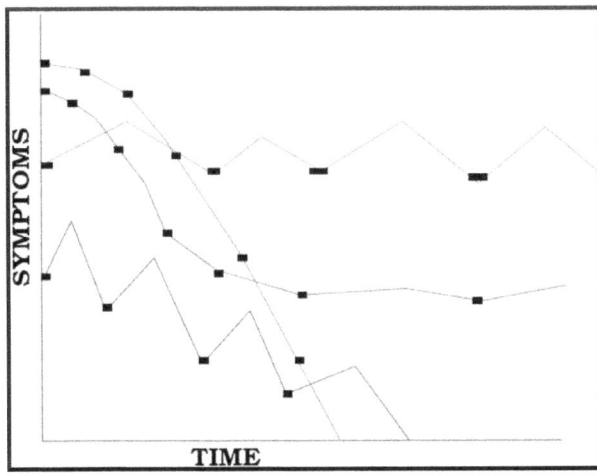

Figure 9.1

9.5   Assessing the response

Deciding when to repeat treatment is important but not always easy. In principle you should give a treatment and observe its effect, repeating it when the initial improvement begins to wear off. But the response to any particular treatment is difficult to predict so timing the repeat accurately may not be practicable. Sometimes one treatment in a sequence may have a lot of effect while the next one does little or nothing. The over-all pattern of response is what you should attend to.

Uncertainties like these make it necessary to treat a lot of patients in order to build up your clinical judgement of when to repeat the treatment. But a good rule to follow is: if in doubt, hold off and observe.

9.6   The longer term

There are several different courses that events may take over a longer period, ranging from complete remission through the 'plateau' (9.6.2) to relapse and the need for 'top-ups'.

9.6.1   Complete remission

In this case there is progressive improvement after each treatment until the symptoms vanish completely. You can now discharge the patient with the instruction to contact you if there is any recurrence in the future.

9.6.2   The 'plateau'

Improvement occurs progressively with repeated treatments until a certain point and then stops in spite of further treatments. This is the 'plateau'.The patient is now better than before starting acupuncture but not completely cured. This is quite a common situation and patients are usually happy to accept it and stop treatment.

If patients are left without further treatment at this stage there may be a slow but progressive improvement in their condition over the next several weeks or months. It is sometimes a good idea to stop treatment for a few weeks to see if this happens.

9.6.3   'Top-up' treatments

In chronic conditions such as osteoarthritis and headaches, 'top-up' treatments are required at intervals, which may be 8 weeks, 12 weeks, or longer. At these times only one or occasionally two treatments are normally needed, even if the initial course to achieve a remission was longer. But what happens if top-ups are needed more frequently—if the treatment works but the effect is brief?

9.6.4   Brief response

There are some patients who respond well to acupuncture but whose improvement doesn't last, even after a full course of treatment. This is a difficult situation. When I first started practising acupuncture I used to get such patients into the wards for more intensive, even daily, treatment, but that never worked. Short remissions are the patient's response pattern in these cases and cannot be changed.

9.6.4.1 Possible solutions for brief response

See if there are any remediable causes for the symptoms, such as faulty footwear or unsatisfactory position at work or when driving. If not, explain the situation to the patient and decide between you what to do. It is possible theoretically to continue acupuncture at frequent intervals but this is usually unsatisfactory. As a rule of thumb I tend to regard eight weeks as the minimum acceptable remission period for continuing treatment.

One possibility in selected cases is for the patient themselves or a relative to give the treatment. This is discussed further in Chapter 23.

9.6.5 No improvement at all

There will always be some patients who fail to respond, either because they are suffering from a condition for which acupuncture is unlikely to work or because they are non-reactors (7.4). So how do you decide when this point has been reached? In other words, when to give up?

9.6.5.1 When to give up

In most cases you should probably give up if there has been no response at all to treatment after two or three sessions. This doesn't mean that there should have been a very good response by this time but *something* should have happened. In very rare cases it may be worth going on longer but usually the law of diminishing returns means that acupuncture is not going to work for this patient.

9.7 Treatment intervals

The timing of treatments is different for chronic and acute conditions.

9.7.1   Chronic conditions

For chronic disorders it is usually convenient to bring patients back a week after their first treatment. Seeing them sooner would not leave long enough for assessment of their response to the initial treatment. But this not a fixed rule and sometimes the second treatment might be given after four or five days.

9.7.2   Acute conditions

In acute disorders the timescale can be shorter. so that patients are treated twice in the first week, for example, and perhaps twice again in the second week. Even so, don't treat patients daily; this would be too often.

9.8   Unexplained symptom changes

Occasionally patients respond well to acupuncture for many months or even years but then cease to do so. I think this is a danger sign; it may indicate that there has been some change in their general state of health unconnected with what they were receiving acupuncture for. In one case, for example, it signalled the development of a malignancy.

A change of this kind should prompt you to review the case and ask the patient if there have been any changes in their general health, such as new pain, unexplained rweight loss, change in appetite or bowel habit, or a drop in energy level.

9.9   Aggravations

This term refers to temporary worsening of the symptoms. The frequency of aggravations is difficult to estimate but my impression is that they are quite common, at least if one takes minor ones into account. My acupuncture web page gets frequent hits from patients inquiring about various adverse effects, especially pain, that they have experienced after acupuncture.

Aggravations may be followed by an improvement. Patients who experience them regularly usually take them in their stride and accept them as part of the process.

9.9.1 Managing aggravations

Most patients who experience aggravations simply require reassurance. Tell them that this does happen from time to time and should pass off quite quickly. You can also say that it may be possible to prevent or reduce the aggravation in future by using more gentle treatment. And say that there is a good chance of an improvement when the aggravation wears off.

9.9.1.1 Two things to look out for.

▷ Make sure that the patient has not stopped analgesic treatment without being told to do so. This can give the appearance of an aggravation although it is not really the case.

▷ If the aggravation seems to be severe or persistent get the patient back and review the situation. Perhaps there has been an infection or a haematoma, for example.

9.9.1.2 Terminating an aggravation

If an aggravation is severe but there is no obvious cause for it you can usually stop it by repeating the original treatment *very lightly*, with insertion lasting not more than a second or two. In practice this is very seldom necessary provided patients are not over-treated in the first place.

Chapter 10

Using the Needles

Chapter Outline

- ▷ Needle sizes and types

- ▷ Depths and angles

- ▷ Holding the needle

- ▷ The needling process in detail

- ▷ Stimulation methods

- ▷ Assessing the response

- ▷ Minimalist technique

- ▷ Periosteal technique

- ▷ Guide tubes

- ▷ Avoiding needlestick injury

- ▷ Avoiding forgetting needles

10.1   Introduction

Acupuncture is a manual technique that requires practice. Here I describe the methods of inserting and manipulating the needles and also indicate some ways of avoiding mishaps.

10.2   Good needling technique

The success of treatment largely depends on two things: the ability to insert the needles quickly and easily and alertness to the effects that needling is having on the patient. These things are often more important than deciding exactly where to needle.

10.3   Acupuncture needles

The needles are made of steel, with handles consisting either of wound wire or plastic. It makes no difference what kind of handle is supplied, except for electro-acupuncture, for which plastic handles are obviously unsuitable.

10.4   Needle sizes

Acupuncture needles come in many different lengths and thicknesses but only a few are widely used.

10.4.1   Medium length needles

These are usually 30 mm in length, 0.30 mm in diameter. Most of your work will be done with them.

10.4.2   Short needles

These are 15 mm in length, 0.20 mm in diameter. They are useful when shallow insertion or gentle stimulation is required.

10.4.3   Long needles

These are 50 mm in length and either 0.30 mm or 0.35 mm in diameter. They are used when deep penetration is required—for example in the lower back.

Still longer needles exist but are seldom used.

10.4.4   Other needles

Very short needles are used in certain types of acupuncture such as ear acupuncture (auriculotherapy). Often these resemble small drawing pins made of wire, which penetrate only about 1 mm. I shall not discuss these further here.

> Throughout the book I offer suggestions for the size of needle to use in different situations. Please understand that these are just that—suggestions. Always use your common sense about which needles to choose, keeping safety in mind. And if I suggest a short (15mm) needle and you don't have one to hand, use a longer needle and don't put it in too far!

10.5   Types of needles

Needles may be either Japanese-style or Chinese-style. The Japanese-style needles tend to be of better-quality steel and are coated with silicon, which makes them easier to insert. It is best always to get this kind of needle. Note that the Japanese-style needles are not necessarily made in Japan; they may come from China or Korea, for example.

Whichever kind you get, make sure that they come in separable individual packets (blisters) with one needle per blister. Some needles are supplied in a large sheet, with perhaps ten or twelve needles together under one cover. This leads to a lot of waste because it is difficult to extract one needle without opening and de-sterilising others as well.

10.5.1   Guide tubes

Needles may be supplied with a guide tube. I discussed this below (10.16).

10.6   The question of depth

There are four possibilities as regards depth.

- ▷ ~~Intradermal~~

- ▷ Subcutaneous

- ▷ Intramuscular

- ▷ Periosteal.

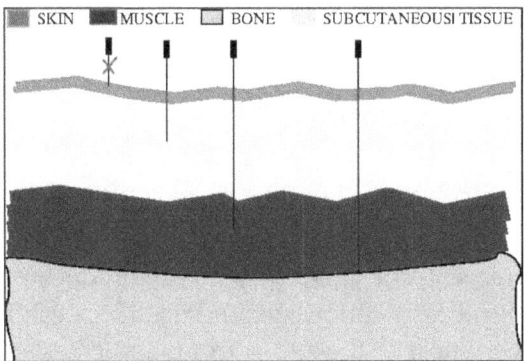

Figure 10.1

10.6.1   Intradermal

This means into the skin but not through it. Avoid this kind of insertion: it is both painful and ineffective.

10.6.2   Subcutaneous

Needles may be inserted into the subcutaneous tissue, whose depth varies according to location and the amount of subcutaneous fat the patient has. Subcutaneous needling is safe; use it if there is any risk of damaging important underlying structures. Contrary to what is sometimes said it can certainly be effective.

10.6.3   Intramuscular

If you want to insert a needle into a muscle trigger point you must obviously needle the muscle concerned. Sometimes there are safety considerations which prevent this (for example the intercostal muscles should *not* be needled for fear of causing pneumothorax). Whether it is always necessary to needle into trigger points directly is also uncertain (5.6.1.1).

.

10.6.4   Periosteal

Two considerations need to be kept in mind when needling the periosteum.

> ▷ Are you traversing any important structures on your way to the periosteum?

> ▷ Are you in danger of entering a joint cavity?

Provided the answer to both questions is no, periosteal acupuncture should be safe.

Periosteal needling is not a precise technique; the periosteum doesn't seem to have accurate central representation and the needle doesn't have to be placed at or even very near to the site of trouble, such as an arthritic joint. How far away it can be placed without losing clinical effectiveness is unclear.

10.7   Needle angle

The choice of angle, vertical or oblique, is largely a matter of common sense, again taking safety into consideration.

10.7.1   Oblique insertion

An oblique angle should be used if there are important underlying structures—for example, in the chest or abdomen.

It is also often better to use an oblique angle when needling a fusiform muscle, since by doing so the needle traverses more muscle tissue and is therefore more likely to hit a TP (assuming that this is what you wish to do).

When needling obliquely in a limb it makes no difference whether the needle is directed proximally or distally. (Why should it?)

10.7.2   Vertical insertion

A vertical angle is appropriate for periosteal needling because in that case you want to reach the periosteum as quickly as possible.

10.8   Holding the needle

Steady your hand by resting your *middle* and *ring fingers* on the patient's skin. **This is important.**

Hold the needle by the handle between your index finger and thumb with the needle **in line with** your index finger. If you then flex and extend finger and thumb the needle will move forwards and penetrate the skin quickly and easily.

Insertion may be difficult if the skin is tough, as it sometimes is, or if the needle is flexible. In that case an alternative technique is available.

10.8.1   Alternative technique

In this technique the base of the needle is pressing against the first phalanx orf your index finger and your finger and thumb squeeze the shaft of the needle above the part which will enter the patient's skin (Figure 10.2). Insert the needle by flexing your wrist.

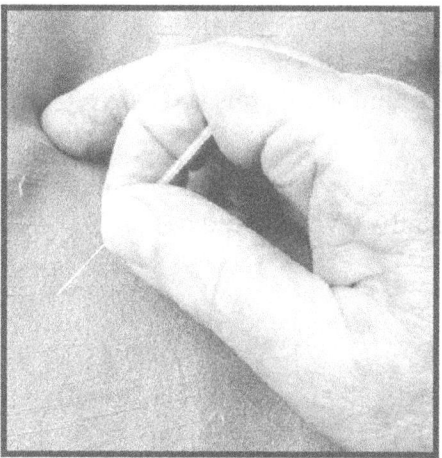

Figure 10.2

This technique has two advantages and one disadvantage.

> ▷ Slipping cannot happen (because the needle base is pressed against your index finger). This makes it easier to penetrate tough skin.

> ▷ The needle is made more rigid, preventing excessive bending and again making skin penetration easier. It can also make periosteal pecking more effective.

> ▷ On the other hand, the proximal part of the needle is touched by the operator's fingers. The technique should therefore be used when deep penetration is not required, such as for periosteal needling when the bone is near the surface (e.g. in the face or a digit).

10.9   Stimulation methods

There are two methods of stimulating the needles manually, twisting and 'pecking'.

10.9.1   Twisting

Rotate the needle by twisting it rapidly back and forth in both directions (not continuously in one direction; the technique is like winding a watch, not turning a screw). This is the standard technique of stimulating soft tissues. Use about a 15 or 20 degree rotation at first, with three or four rotations.

10.9.2 'Pecking'

Lift and press down the needle a few times without taking the point out of the skin. This is the standard method of stimulating the periosteum. The needle point strikes the periosteum at the end of each 'stroke' The treatment is quick: three or found pecks are used in a space of a few seconds.

> *Mnemonic:* Use **T**wisting for soft **T**issues, **P**ecking for **P**eriosteum.

10.10 The needling process in more detail

Here I describe the process as a series of steps; this is simply for clarity. In practice the steps follow one another smoothly, without interruption. (For needling using a guide tube see 10.14.)

10.10.1 Preliminaries

▷ Cover any wound in your hands with a waterproof dressing, for both your and the patient's protection.

▷ Wash your hands thoroughly with soap and water or wipe them carefully with an alcohol-based solution. Note that alcohol will not be effective if your hands are not clean, and it may not work for some viruses, such as norovirus.

10.10.2 Inserting the needles

Warn patients before the needle is inserted, otherwise they may 'jump' and perhaps bend the needle. I usually say "Needle going in now ...".

The technique differs according to whether soft tissue or periosteal needling is to be done. I describe these separately.

10.11   Soft-tissue needling

▷ Stretch the skin with the free hand (usually the left). The aim is to make the skin as taut as possible. To enhance this, notice that the skin stretches differently in various parts of the body, according to how most of the elastic fibres run. The skin should ideally be stretched in the opposite direction to that of the fibres to increase the tautness.

▷ In some places, for example over the lungs or abdomen in a thin patient, it may be necessary to pick up a fold of skin and insert the needle horizontally. Do this if there is a concern about safety, but this technique is usually more painful than the standard one so don't use it routinely.

▷ Rest the needle tip on the skin; don't 'dart' it in from a distance.

▷ Penetrate the skin quickly, without hesitating. If you hesitate in a misplaced attempt to avoid causing unnecessary pain, it will actually hurt more than if you get through the skin fast. In some cases it can be surprisingly tough—for example, on the back of the neck in elderly men.

▷ Continue to advance the needle to the desired depth, which you have decided in advance (10.5). As you do this you will get manual feedback from the needle. For example, you may feel it penetrating various layers of fascia, and in muscle it may be gripped to cause a resistance to rotation ('muscle grasp').

▷ Stimulate (twist) the needle three or four times; then stop and ask the patient "What do you feel?" If they reply "It's hurting," ask whether it only hurt briefly as it went in or is still hurting. If it is still hurting appreciably, remove it.

From this point the duration of needling and the amount of stimulation will depend on what the patient reports. Assessing these effects is very important.

10.12   Assessing the response

Apart from pain there are three things to noter: local sensations, remote sensations, and generalised effects. If necessary prompt the patient for information about these.

▷ "Do you feel anything round the needle?"

▷ "Do you feel anything anywhere else?"

▷ "Do you feel any different in yourself?"

If a patient asks "What am I supposed to feel?", don't offer any suggestions but simply ask if there are changes of any kind.

10.12.1   Local sensations and effects

The local sensations are mainly those referred to earlier as *de qi* (7.2.2). In addition there may be visual effects such as flushing of the skin round the site of needle insertion, which indicates a local reaction with release of substances such as calcitonin gene-related peptide, substance P, and histamine (which will cause local itching).

10.12.2   Remote sensations and effects

These might be, for example, a feeling of warmth in the chest or neck when a needle is inserted in the foot. There may also be flushing in the chest or throat. There are many other possibilities. I am constantly surprised by the effects that I see when inserting needles in patients.

10.12.3   Generalised sensations

The generalised sensations may be relaxation or other effects, such as euphoria, drowsiness, or a 'spaced-out' feeling. Patients may say that it is as if they have had a few drinks or, if they are familiar with the sensation, have taken cannabis. Occasionally there are even stranger effects. Some patients remark spontaneously that everything looks brighter. Rarely, patients seem unable to speak as long as the needles are in place. They may laugh or cry, apparently without knowing why.

> Success in acupuncture depends largely on alertness to the patient's response and on getting the amount of stimulus right. It is easy to do too much and difficult to do too little.

10.13   Withdrawing the needle

Now remove the needle and dispose of it safely. **Do this one-handed to reduce the risk of needlestick injury** (see 10.16).

> The total duration of needling is about 2 minutes maximum per needle and often much less. The reasons for using brief needling are described in 8.1.3.

10.13.1   Locked needle

Occasionally a needle becomes locked in the muscle and cannot be withdrawn. Books often advise inserting another needle close by to relax the muscle, but I assume that the needle is doing something useful in these cases so I simply wait until it can be removed without too much pulling (a gentle twist may help to free it). In this case the needle may remain in place for longer than usual.

Don't confuse this kind of locking, which only happens in large muscles (mainly the gluteals), with adherence to the needle due to the elastic fibres in the skin. True muscle locking is actually quite uncommon.

10.14   Minimalist technique

This important method was first described by Felix Mann, who termed it 'micro-acupuncture'. It is similar to the technique just described except that a short fine needle (15 mm) is used and the insertion is extremely brief, perhaps as little as 1 second.

Intuitively it might seem unlikely that such minimalist treatment could have any effect but in fact it does, and there are some patients who will only respond to this kind of treatment. It should always be used on children and ultra-strong reactors.

10.15   Periosteal technique

This is quite different from soft-tissue needling. Penetrate the skin quickly as usual and then advance the needle until it reaches the periosteum. There should be an unmistakable hard end point, like hitting wood. Peck the periosteum firmly three or four time quickly and then withdraw the needle. The whole treatment takes only a few seconds.

A note on terminology

As Felix Mann has remarked the periosteum is actually quite thin and the technique could also be described as bone acupuncture, but the periosteum has a richer nerve supply than bone so it is probably this tissue that is responsible for the therapeutic response. The term 'periosteal acupuncture' is now widely used by modernists.

The pain caused by periosteal acupuncture is very variable. Some patients feel a lot, others nothing at all. The typical periosteal sensation is a deep dull ache, which is somewhat unpleasant but not agonising.

The amount of sensation elicited by periosteal acupuncture is unimportant for its effect. So there is no need to ask for the patient's sensations, though you may do so as a matter of courtesy. What we are really interested in is the response of the symptoms (typically, joint pain) over the next day or two.

10.16   Guide tube technique

These needles are housed in a plastic tube, with just the top of the handle projecting about 3 mm. They may be held in place in the tubes in various ways, perhaps by a little wedge or stopper at one end or by a touch of glue at the top.

The way to use these needles depends on exactly how they are held in the tube, but in any case, you place the end of the tube on the skin at the site to be needled and press the the top of the needle firmly to drive it through the skin. Then remove the tube and advance the needle further to the desired depth.

10.16.1   Advantages of using guide tubes

▷ It makes it easier to penetrate the skin for beginners.

▷ Insertion may be less painful because there is a slight local anaesthetic effect due to distraction of the nervous system by the pressure of the tube.

▷ It might appear to be more sterile although the needle still has to be advanced by hand so this effect is probably unreal.

10.16.2   Disadvantages of using guide tubes

▷ The procedure is rather awkward and takes longer than not using guide tubes so many experienced acupuncturists, including me, dislike them.

▷ If you become dependent on using the tubes you feel at a loss if, for some reason, you are confronted with needles that lack them.

▷ Some people are tempted to replace the needle in the sheath after use, which is dangerous.

10.16.3   Verdict on guide tubes

There is no reason not to use them if you want to. They do make insertion easier for beginners, particularly when using long needles. But I favour learning to use needles without the tubes since once you are used to that method it is quicker and easier to perform. It is also slightly cheaper to buy needles without tubes.

10.17 Avoiding needle-stick injury

These are the precautions I take to minimise the risk of needlestick injury.

10.17.1 Use each needle once only

It is possible to insert a needle more than once in the same patient, but each time you spread the skin to make it taut there is a chance that you will needle your free hand. So my invariable rule is to dispose of any needle that has penetrated the skin, even only a millimetre.

10.17.2 Dispose of needles immediately

There is a tendency to keep a needle in your hand as you are writing up your notes or chatting to the patient or to someone else in the clinic, but this is likely to result in sticking the needle accidentally into yourself, or, even worse, into someone else. So get rid of it at once.

When disposing of the needle, hold it horizontally and drop it in the bin (at least if it has a full-size aperture). If the needle is put in point-first it can easily catch on the edge and bounce out. Make sure it has really gone in.

10.17.3 Take needles out with one hand

For some reason beginners in acupuncture find this psychologically difficult to do, arnd books often advise you to press the skin as you take the needle out. If you do this you will certainly stab yourself sooner or later as the needle comes out, and this is of course a needlestick injury with all that that entails. I therefore strongly advise taking needles out using only one hand. Keep your free hand well away from the site of insertion. This will feel a little strange at first but after a short time it will become completely automatic.

10.18 To avoid forgetting needles

It is easy to forget a needle. To reduce the risk do the following.

▷ Record the number of needles used.

▷ Count them in and count them out.

▷ Keep the empty packets; the numbers of needles and packets should match.

▷ As a rule, withdraw each needle before putting in another (but see the next section).

The last point has two advantages.

▷ It is impossible to leave a needle behind.

▷ If there is a particularly strong effect from one insertion you are sure which it is.

But there are occasions when I break the one-needle-at-a-time rule.

Using multiple needles

Although I usually use only a few needles and sometimes only one, I do sometimes insert more. If I am using more than four I put them all in together, otherwise it takes too long. Putting in more needles entails a greater risk that one will be left behind, so I run my hand quickly over the area after removing all the needles to make sure they are all out.

When putting in a number of needles at the same time I warn patients that an occasional insertion may be painful. I ask them to let me know if that happens so that I can remove the offending needle immediately.

Some of the illustrations in this book depict a very long 'needle' (actually, a bicycle spoke!). This is simply to indicate the direction of insertion more clearly than a small needle would do.

Chapter 11

Three Types of Spinal Pain

Chapter Outline

  ▷ **A simple classification of spinal pain**

  ▷ **Response rates to acupuncture**

11.1   Introduction

Pain in the neck, thoracic spine, or lumbar spine is a common complaint that acupuncturists have to deal with. The causes of spinal pain are complicated and there is much uncertainty about them. Modern methods of imaging such as magnetic resonance imaging and computed tomography can provide information about what is happening, but even these can be misleading and they are not often used for acute or fairly mild chronic pain.

For practical purposes we need a simple way of classifying neck and back pain. I use the following three-fold scheme, which affords at least a tentative basis for assessing the likelihood that acupuncture will be helpful.

11.2   Type A pain

This is probably the commonest type of spinal pain. It is deep, dull, and aching in character and is poorly localised. It may be felt only in the back or may radiate down a limb. The mechanism of this kind of pain is not well understood. It may respond to acupuncture.

11.3   Type B pain

This is almost the opposite of Type A pain. It is felt locally in the midline, and seems to be in the superficial tissues overlying the vertebral spines. There is no radiation. It most commonly occurs at the lower part of the neck or in the upper thoracic region. It often responds well to acupuncture (13.2.6).

11.4   Type C pain

This type of pain appears to be due to pressure on a nerve root; for example, by a prolapsed intervertebral disk. There is usually a dermatome distribution of referred pain, and there may be either increased or decreased sensitivity or electric-shock-like sensations in the affected limb. There may also be motor changes, with loss of strength and absent reflexes. The chance that acupuncture will work in this case is reduced, because needling the periphery will not eliminate pressure on a spinal root. But it is still worth trying acupuncture a few times because there is always room for doubt in assessing the pathology.

For example, a man with neck pain radiating to his arm was seen by a neurosurgeon who advised operation, although he warned the patient that there was a small chance that he might end up tetraplegic. The man therefore sought an alternative to surgery. After two sessions of acupuncture in which his articular pillar was needled periosteally (13.4.4) his symptoms were completely relieved without subsequent relapse.

Chapter 12

Generalised Effects

Chapter Outline

▷ Generalised (central) effects

▷ LR3 vs LI4

▷ Avoiding over-precision in needling LR3

▷ Practical aspects of needling LR3

▷ Clinical indications

12.1 Introduction

In many patients acupuncture produces subjective changes that are presumably mediated by mechanisms in the brain. This can occur with needles inserted anywhere and is particularly likely to happen to strong reactors. But some areas of the body seem to be more capable of eliciting such responses than others. The feet and the hands are most important in this respect.

12.2   LR3 vs LI4

Two classic sites are commonly used to produce central effects of this kind. One is on the hand, in the web between the thumb and first finger (adductor pollicis), and is known as Large Intestine 4 (LI4). The other is on the dorsum of the foot between the first and second metatarsals (first dorsal interosseous); this is Liver 3 (LR3). I prefer to avoid LI4, both because it is probably less effective than LR3 and because it is potentially dangerous (15.6). I therefore always use LR3 for this purpose.

12.2.1   Exact location of LR3?

For reasons I explained in Chapter 1 I am sceptical about the real existence of any classic points, including LR3, so I certainly don't want to give the impression that you must needle this site with the accuracy that a traditionalist would demand. As usual it is a question of an area not a 'point'.

But how big an area? Would it make any difference if you used a different interspace? Probably not; probably anywhere on the dorsum of the foot would be equally effective. In a strong reactor anywhere below the knee might be just as good. Incidentally, this is the likely explanation for claims for widespread effects from needling classic sites in the anterolateral leg (20.4).

In spite of these reservations I do use the label LR3 for the treatment as a convenient shorthand description. But please keep in mind that I am really referring to an area of uncertain size in the foot rather than to a small 'point'.

12.3   Needling technique

Dorsiflex the foot and palpate the dorsum to identify the first and second metatarsals. This is usually easy to do but may be difficult in a fat foot. Hold the forefoot with your fingers on the sole and your thumb on the dorsum stretching the skin. Keep the tension up until you withdraw the needle since it will move unpredictably if you remove your thumb too soon.

Insert the needle in the middle of the interspace . For some reason traditional books show LR3 to be a little more proximal than this, which would put it right over the dorsalis pedis artery. I have not heard of anyone damaging this vessel but I prefer to avoid the risk. You can use either a 30mm or a 15mm needle.

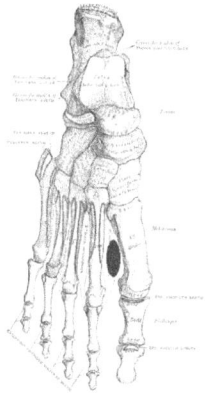

Figure 12.1

Once the needle is in place start to stimulate it by twisting, stopping frequently to get feedback from the patient. We are particularly interested in any remote (often chest or neck) or generalised (central) effects there may be.

If, as occasionally happens, there is a very marked response almost at once, take the needle out immediately.

12.4 Side differences

There is usually a difference in reactivity between the two sides. In about 70 per cent of people the left side is more reactive than the right. (This has nothing to do with being left-handed or right-handed.) In the remainder either there is no difference or the right is more reactive than the left. So start by needling the right side in case the patient turns out to be a strong reactor. If nothing much happens needle the left side in the same way. But if you get a very strong response to needling the right side don't do any more.

12.5 Indications for using LR3

In principle LR3 can be tried for any disorder and it will sometimes work for quite unlikely indications in a strong reactor. It can also be used to enhance the effect of needling other sites.

Here are some disorders for which I would normally try LR3 in the first instance as most likely to work.

12.5.1 Chronic urticaria

This quite often responds to acupuncture although the effect may be short-lived. If so it may be possible to teach the patient to treat themselves, perhaps once a week (see Chapter 23).

12.5.2 Polymorphic light eruption

This is a disorder in which patients develop a rash whenever they go out in sunlight. It can be due to medication or to contact with various plants but it may develop for no known reason. The response to acupuncture is usually good and patients who suffer from polymorphic light eruption are often strong reactors to acupuncture.

12.5.3 Hot flushes

These occur in women around the time of the menopause and also in men who are receiving hormonal treatment for prostate cancer. There is often a good response to acupuncture; again, self-treatment may be useful.

12.5.4 Headache

LR3 is often used for headaches (13.5).

12.5.5 Bronchial asthma

Most trials of acupuncture for asthma carried out in the West have shown no effect and I agree that acupuncture generally does not work for this disorder. But just occasionally it produces outstanding results, with 400 per cent increases in peak flow demonstrable immediately after the needles are inserted. The effect in such cases usually lasts for several weeks before beginning to wear off, at which point the acupuncture should be repeated.

Traditional books list numerous points that are supposed to work for asthma but I have not found them to be useful. If anything is going to work it is LR3. If that fails, as it often does, there is no use continuing.

Chapter 13

Head and Neck

Chapter Outline

- ▷ Neck pain

- ▷ Headache

- ▷ Vertigo

- ▷ Temporo-mandibular joint dysfunction

- ▷ Vasomotor rhinitis

- ▷ Trigeminal neuralgia

13.1 Introduction

From a practical point of view I divide the treatments in this region into those that are most easily done with the patient sitting and those that are most easily done with the patient lying down. This is not a rigid division; some practitioners like to do all the treatment with the patient lying down, but I prefer to use the methods described here.

13.1.1   Risk of fainting

When treating patients sitting, consider the possibility that they may faint, particularly on the first occasion. If that seems to be a risk patients should sit sideways on a couch facing away from you; in that way, if they do feel faint they can be brought to a lying position with no danger of their falling to the floor. Remember that the most likely patients to faint are young athletic men!

13.2   TREATMENTS WITH PATIENTS SITTING

13.3   Trapezius sites

There are frequently TPs in trapezius. The commonest one is at the midpoint of the trapezius and is known in the traditional system as Gall Bladder 21 (GB21). Note that this is shown quite large in Figure 13.1, to emphasise that this is just a guide to where to start looking and is not a definite 'point'. TPs may be found anywhere in the trapezius and more than one may be needled.

There is quite often a TP at the junction between the neck and the shoulder, in levator scapulae. This is not a classic point but I shall call it GB20.5 because it is half-way between GB21 and GB20.

GB20, at the base of the skull, is most easily treated with the patient prone; it is discussed below (13.4.6).

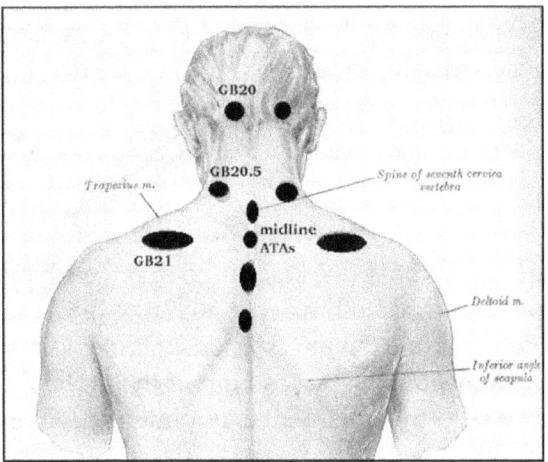

Figure 13.1

13.3.1   Pneumothorax risk

 Needling GB21 or other areas in the trapezius risks causing pneumothorax. The apex of the lung rises 2 to 3 cm above the medial third of the clavicle and has been injured a number of times by acupuncturists in the West.

13.3.2   Needling technique

To avoid this danger **lift the trapezius up with your free hand and direct the needle forwards from behind**. This means that in a sitting patient the needle will be parallel to the floor. If the patient is lying down the needle will be directed downwards towards the couch.

Keep the muscle lifted up until you withdraw the needle. If you let it go too soon the needle is likely to be drawn downwards by the rebounding tissues.

13.3.3   'GB20.5'

This is needled in the same way as GB21. This applies to any other sites in trapezius that may be needled because they are tender.

13.3.4   Use bilateral needling

In most cases you should needle both sides, but do more work (more stimulation and possibly more needles) on the side which has the greater degree of tenderness. But very sensitive patients should be treated on one side only.

13.3.5   Midline area

Some patients experience superficial pain over the spine in the midline (Type B pain—see 11.3). This can be treated by local needling over the spine. The needle will usually penetrate the interspinous ligament. In a very thin patient it is possible to needle the periosteum of the vertebral spines, but this is not necessary and the treatment is not intended to be periosteal.

To avoid any (remote) danger of injuring the spinal cord, use short (15mm) needles if available. Because of the angle of the spines, which slope downwards like the tiles on a roof, direct the needles slightly caudally at about 45 degrees.

Insert the needles at approximately 2cm intervals, using as many as necessary to cover the whole painful area. In practice this means that about 4 or 5 needles are inserted in most cases. This is a good treatment with a high success rate.

13.4 Psychological disorders

On the whole acupuncture is not outstandingly successful in the treatment of psychological disorders whereas it can work well when there is definite objective pathology, as in osteoarthritis. But there is a psychological component in many clinical problems and acupuncture does have some effect in mild or moderate anxiety and depression.

LR3 is often recommended in such cases but I have had more success in using the trapezius (GB21). The rationale for this is that tension in the neck and shoulder muscles is a common feature in these cases. The anxiety or depression can cause tense muscles but the reverse is also true, and relaxing the muscles can help the psychological symptoms.

This no doubt explains why acupuncture cookbooks often include GB21 in their 'recipes'.

13.5   TREATMENTS WITH PATIENTS LYING DOWN

13.6 Trigger zones in posterior triangle

Many patients have TPs in semispinalis capitis and cervicis or in splenius capitis and cervicis. Although these can be needled in a sitting patient it is easier to do so with the patient lying on their side.

13.6.1 Needling technique

The patient lies on his or her side with the affected side uppermost. Place a pillow under the patient's neck to flex it laterally. Palpate the muscles in the posterior triangle and identify any TPs. Insert the needle at right angles to the skin, directing it towards the cervical spine.

Judge the depth of insertion according to the patient's size. In most cases use a 30mm needle inserted to about half its length.

13.6.2   Condition treated by needling posterior triangle

  ▷ Neck pain, with or without radiation

  ▷ Carpal tunnel syndrome (mild or moderate severity)

  ▷ Autonomic dysfunction in hands

Carpal tunnel syndrome can be helped in about a third of patients. This is about the same as the spontaneous recovery rate in this condition. This suggests that acupuncture can sometimes speed up spontaneous recovery—something also seen in other disorders.

Vertigo arising from abnormal proprioception in neck

The last-mentioned condition (vertigo arising from the neck) is somewhat controversial, but I think it probably does occur. Typically, the patients, who are middle-aged or elderly, complain of dizziness when turning the head. In at least some cases it appears to be due to abnormal proprioceptive information from the cervical joints or the neck muscles. In other cases it is ascribedkkkkkkkkk to impaired circulation in the neck vasculature, in which case acupuncture is less likely to work. Since it is not easy to distinguish between these possible causes clinically, acupuncture is always worth trying.

13.6.3   Periosteal technique

A variant treatment that can be used *in thin patients only* is to needle the articular pillar of the cervical vertebrae periosteally. This area is situated *behind* the transverse processes.

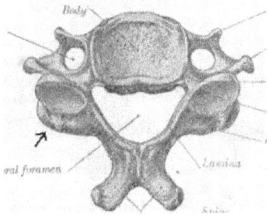

Figure 13.2

Palpating it has been described (by Hywel Watkin) as like feeling an armadillo.

Figure 13.3

A 15mm needle should usually be long enough. In a thin patient the periosteum is near the surface and will be reached almost immediately. If you don't reach the periosteum after inserting the needle to about 1 cm don't insist.

This technique was first described by Mann, who found that it could affect almost any condition in the upper half of the body. While I agree about its effectiveness I think that in many cases simply needling the overlying muscles has much the same effect and is technically easier to do.

13.7   Sternocleidomastoid

This is not a muscle that I find I need to treat very frequently, but it is worth keeping in mind for pain syndromes in the area. I have seen trigger points near its insertion give rise to referred pain to the root of the tongue.

13.7.0.1 Needling technique

With the patient supine, the head is rotated to the side opposite the muscle to be treated and raised slightly. This makes the muscle stand out, so that you can grip it between finger and thumb and palpate it for TPs. If one is found, insert a 15mm needle into the tender area or areas from lateral to medial. **Avoid the external jugular vein** (usually easily seen).

13.8 Occipital site (GB20)

This site is important for headaches (discussed below) and sensations can radiate to the orbit on the same side. It can be used to treat a unilateral watering eye.

13.8.1 Needling technique

Place the patient prone and palpate the skull. A common mistake is to start too low. To avoid this your fingers should start on the skull and then slide downwards to where the skull slopes away. The needling site is between the insertions of sternocleidomastoid and trapezius.

Insert a 30mm needle and direct it towards the opposite eye. The periosteum will be encountered after the needle has been inserted to about half its length or a little more.

Tap the periosteum quickly three or four times. The patient will hear this via bone conduction.

The treatment can also be done with the patient sitting but it is more difficult to do so. Although it is not usually very painful, this is a reactive site and is quite likely to make the patient faint if he or she is sitting. This is an additional reason for having the patient lying down.

13.9 Headache treatment

Headache is one of the commonest conditions treated with acupuncture and the success rate is high (more than 70 per cent). Both tension headaches and migraine, with or without aura, often respond (25.2.3, 25.2.4).

Acupuncture works mainly as a preventive in headaches. It reduces either the severity or the frequency of the headaches, or—ideally—both. It can sometimes relieve a headache that is already present but it never stops the progress of a migraine aura.

13.9.1 Two approaches to treatment

There are two ways of treating headaches.

▷ Generalised (central) stimulation via LR3.

▷ Local treatment via the neck, using GB21 and GB20 bilaterally. Probably only GB20 is really required in most cases but I usually needle all four sites.

13.9.1.1 General or local: how to choose

There is no infallible method of choosing which treatment to use so try both at least twice before giving up. Probably more patients respond to the neck approach. There are some tentative guidelines.

▷ If there are active TPs in the neck it is logical to start with those.

▷ If there is a large element of psychological tension it may be better to start with the neck, since treating this region may reduce tension (13.3).

▷ If there is a migraine aura (usually but not always visual) it is probably better to begin with LR3.

▷ If there is a clear history that headaches are set off by certain foods or alcohol LR3 is more likely to work.

It might seem logical to try both treatments simultaneously in resistant cases but I have never found this to work. As usual in acupuncture, doing more is hardly ever the answer.

13.9.2 Aggravations

Aggravations seem to be particularly common in headache sufferers. The treatment may start a headache later in the day so you should be careful when giving the first treatment; patients should be warned of the possibility of aggravation and if they have something important to do that day it is better to postpone treatment.

13.9.3 Treatment patterns

A typical course of treatment might be as follows.

 ▷ After the first treatment, either to the neck or LR3, the patient returns after, say, two weeks. (The interval depends on the frequency of headaches; if they are occurring more than once a week the second appointment might be a week later.)

 ▷ If there has been an improvement repeat the treatment and bring the patient back after another two weeks.

 ▷ If there has been no improvement either give one more treatment at the same sites or change to the alternative approach.

Once an improvement has occurred the intervals between treatments are lengthened progressively. After perhaps six sessions the patient will usually be able to go several months between treatments. But acupuncture does not usually provide permanent relief and most patients will require repeat treatments perhaps four times a year. If stress is a major factor more treatments may be required at times of psychological difficulty.

13.10   Migraine in children

Migraine in children may respond exceptionally well. A quick insertion at LR3, without manual stimulation, may give relief lasting for several months, even in children who are getting migraine a couple of times a week and are missing large amounts of schooling.

13.11 Three types of resistant headache

▷ Headaches occurring at the time of menstruation are often resistant to acupuncture although those occurring at other times may clear up. (The same is true of medication treatment.)

▷ Cluster headaches are extremely severe headaches, typically centred over one eye and occurring more often in men (unlike migraine, which is commoner in women). They often occur daily in 'clusters' for about six weeks and then go away for some months, although there are many variations. They don't normally respond to acupuncture.

▷ Constant daily headaches almost never respond (see next section).

13.11.1 Analgesic over-use

A common cause for daily headaches is over-use of analgesics. Patients will say they have taken medication several times a day for months or years. They think they are getting the headaches *in spite* of the medication but in fact they are getting them *because of* it.

The only effective treatment is such cases is withdrawal of the medication. This is best done abruptly, but in some cases that is not possible and gradual withdrawal is necessary. If you are not responsible for the patient's medication ask the GP to supervise this.

13.11.2 Differential diagnosis: cranial arteritis

Cranial arteritis (giant cell arteritis, temporal arteritis) is an important condition because of the **risk of sudden blindness**. The characteristic  symptoms are headache, tenderness mainly though not always in the temporal region, and pain on chewing. If headache develops for the first time in an older patient consider the possibility of cranial arteritis. If you suspect it (and are not the patient's GP) he or she requires urgent referral for consideration of corticosteroid treatment.

13.12   Trigeminal neuralgia

Acupuncture can be helpful in this disorder. The treatment consists in deep needling in the infratemporal fossa.

Needling technique

> ▷ With the patient lying on their side, locate the lower border of the zygomatic arch and the anterior border of the ramus of the mandible, which can be felt, rather indistinctly, through the masseter.

> ▷ Insert a 30mm needle in the angle between the zygomatic arch and the ramus of the mandible, aiming slightly upwards and backwards. Stimulate gently for about 30 seconds. Sometimes the needle will touch the periosteum of the maxilla; this doesn't matter although the treatment is not intended to be periosteal.

> ▷ I find it is better to treat both sides.

> ▷ Sometimes I supplement the treatment with subcutaneous needling along the line of pain distribution.

It generally takes about six treatments to get the maximum remission in this condition. Patients will usually require repeated courses every two or three months, whenever they relapse. Acupuncture may not eliminate the need for medication completely but it usually reduces it.

13.13   Vasomotor rhinitis

In vasomtor rhinitis patients experience frequent sneezing, clear nasal discharge, and nasal irritation. It often responds well to periosteal needling over the frontal and maxillary sinuses (approximately BL2 and ST2 in TCM terms). Use a 15mm needle (less painful) and place a finger at the lower margin of the orbit to make sure the eye is not injured. The same treatment can be used for allergic rhinitis (together with LR3).

13.14   Facial hemispasm (not blepharospasm)

In this disorder patients experience spasmodic contraction of the muscles round the orbit. It can be treated by needling the infraorbital nerve. This is a branch of the maxillary nerve which emerges from the infraorbital foramen and divides into a number of branches that run down over the maxilla. Insert the needle just below the foramen and lightly peck the periosteum in a little arc to catch the nerve branches.

This treatment is effective, at least in the short term, for treating facial hemispasm. Some patients have already found for themselves that pressing this area with an object such as a pencil point can relieve the symptoms temporarily. Although it is interesting that a needle can have this effect, it is not always long-lasting enough to make it worth while using. In one case a patient obtained relief by using TENS, with one of the pads placed over the infraorbital nerve and the other on the ipsilateral trapezius.

13.15   Temporo-mandibular joint dysfunction

This responds fairly well to needling the muscles above and below the joint (temporalis and masseter). If the problem appears to be in the joint itself, try periosteal needling to the posterior part of the zygomatic arch and the ramus of the mandible.

Chapter 14

Shoulder

Chapter Outline

 ▷ **Success rates**

 ▷ **Periosteal treatments**

 ▷ **Soft tissue treatments**

14.1 Introduction

The shoulder is not one of the best regions to treat with acupuncture. The anatomy and pathology of this region are complex and pain in the shoulder may really be referred from elsewhere. The chances of success with acupuncture will vary according to the cause of the pain.

14.2 Shoulder or neck?

Pain in the shoulder may be referred from the neck. A useful pointer to distinguishing its origin is the 'bra-strap divider'. Pain situated proximal to this line is probably coming from the neck; if distal it is probably due to a shoulder problem (rotator cuff or frozen shoulder).

14.3 Frozen shoulder: clinical features

▷ Shoulder stiffness

▷ Severe pain

▷ Loss of active and passive external rotation

▷ Lasts about 30 months on average

▷ Recovery usual but full range of movement may not return

Probably about a third of patients with frozen shoulder respond to some extent to acupuncture, with a temporary diminution of pain and better sleep. The course of the disease is not altered but if the acupuncture has even a temporary effect on the pain it may be worth continuing treatment at intervals until spontaneous resolution occurs. While this is not a brilliant result it is about the same as may be expected from the current conventional use of corticosteroid injections into the joint (26.3.5).

14.4 Acupuncture techniques

As in many regions of the body, acupuncture treatment here includes both soft tissue and periosteal components. Deciding which to use depends largely on whether you think the problem is an intrinsic joint disorder (frozen shoulder) or is arising from the muscles controlling the shoulder (rotator cuff disorder). In the first case you would think of periosteal acupuncture, while the second would prompt a search for trigger points.

14.5 Periosteal sites

These are mainly used for frozen shoulder.

14.5.1   Coracoid process

This is probably the most effective site for intrinsic shoulder pain. In most people it is easily palpable but it can be difficult to feel in fat patients, very muscular patients, and those who habitually hunch their shoulders forwards. In most patients the coracoid is near the surface and the needle will reach the periosteum quite quickly. **Be careful not to injure the subclavian vessels or brachial plexus**.

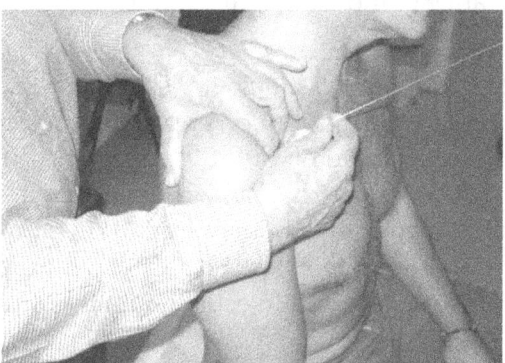

Figure 14.1

14.5.2   Subacromial space (subacromial bursa)

The indication for needling this is a painful arc on abducting the arm. Although it is not strictly a periosteal site it is convenient to describe it under that heading. The needle is inserted into the bursa from the lateral side.

The bursa does not normally communicate with the shoulder joint but it may do so in older patients because of the increasing presence of defects in the capsule with age. In that case there is theoretically a risk of causing a **septic arthritis** so particular care with hand washing is required for this treatment.

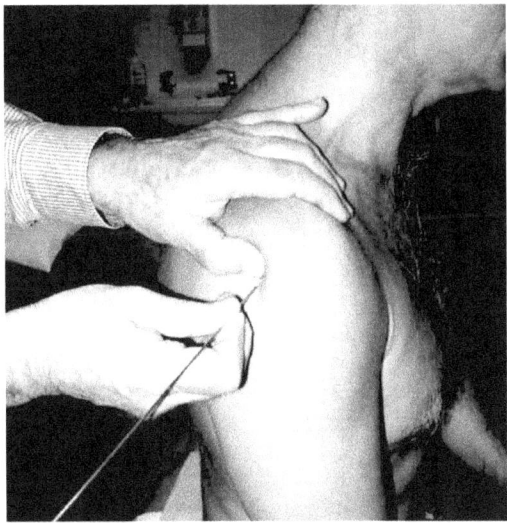

Figure 14.2

14.6 Soft-tissue treatments

These are required for rotator cuff problems.

14.6.1 Axillary folds

Trigger points occur in the anterior or posterior axillary folds (pectoralis major, subscapularis, teres major and minor, latissimus dorsi).

14.6.1.1 Needling technique

▷ To needle the anterior fold safely, have the patient supine, abduct the arm and place your free hand in the axilla with the back of your fingers resting on the patient's ribs. Direct the needle towards your finger tips. **Note that trigger points in this region can cause angina pectoris.** (See 16.3.6 for further details.)

▷ To needle the posterior fold, grip the muscles with your free hand, placing your thumb in the axilla. In a thin patient you can grip the scapula between finger and thumb; this allows you to palpate subscapularis. Needling in this region is directed anteroposteriorly, parallel to the chest wall.

14.6.2   Supraspinatus

Identify the spine of the scapula. By pressing just in front of the spine your fingers are on supraspinatus, with no danger of causing a pneumothorax because the needle is separated from the lung by the floor of the supraspinous fossa. The needle should follow the direction of the fibres towards the tip of the shoulder.

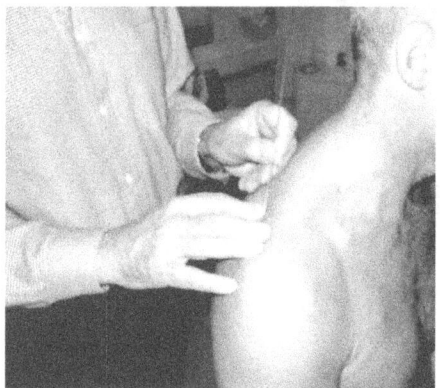

Figure 14.3

14.6.3   Infraspinatus

A trigger point quite often forms in this muscle at about the middle of the scapula. Identify the medial and lateral borders and the apex of the scapula. The needle should follow the fibres of infraspinatus out towards the shoulder. **Avoid periosteal needling of the scapula.** It is unnecessary and there is a small chance that the needle might go through the thin bone in this region and cause a pneumothorax.

14.6.4   The rhomboids

Trigger zones may form close to the attachments of these muscles to the medial border of the scapula. Place the patient prone with the hand resting on the lumbar spine at the back. This will make the medial border of the scapula stand out. A needle may then be inserted parallel to the medial border to catch the attachments of the rhomboids. (Whether the needle points up or down is unimportant.)

Chapter 15

Upper Limb

Chapter Outline

 ▷ **Epicondyltis**

 ▷ **Referred wrist pain**

 ▷ **Wrist sites**

 ▷ **Hand sites**

15.1   Introduction

Acupuncture is effective for many problems in this region apart from epicondylitis, which usually does not do well.

15.2   Epicondylitis

There are two treatments that may be tried.

 ▷ Needle the common tendinous origin. This is the acupuncture equivalent o of corticosteroid injection into the area. There is still much disagreement about the long-term usefulness of injection and I have not found acupuncture to be very effective either. It is also painful.

▷ Look for TPs in the muscles above and below the elbow and needle these; it helps in some cases. Be sure to avoid the vessels and nerve in the middle of the antecubital fossa. I find about a third of patients respond to this.

15.2.1   New treatment for epicondylitis

A new treatment for epicondylitis, unconnected with acupuncture, has recently been described by Bruce Rothschild and appears to have a high success rate (26.3.6). It consists simply in applying a Velcro band 1 inch (2.5 cm) below the epicondyle. If further experience confirms its effectiveness this will presumably become the treatment of choice for this condition.

15.3   Pain referred to the wrist

Most people have a latent TP in extensor digitorum, in the strand of this muscle which extends the middle finger. A twitch can often be elicited here by 'plucking' the muscle like a guitar string, even in people who have no symptoms.

When activated, for example by using the fingers a lot in typing, the TP can refer pain to the dorsum of the wrist. The pain is usually quite localised in the centre of the wrist and can be confused with osteoarthritis if the possibility of referred pain is not considered.

15.3.0.1   Needling technique

Support the forearm with the wrist flexed. Either place the forearm on a cushion or ask the patient to hold it horizontal by placing the other hand under the wrist.

Identify the TP by pressure and insert a 30mm needle obliquely to catch more of the muscle. It is probably necessary to needle this TP accurately, so if you miss it at first, probe the muscle again to try to locate it. When you hit the spot the patient will probably feel sensations travelling down the arm to the site of referred pain.

This is a good treatment which usually gives permanent relief if the site is identified accurately.

15.4 Anti-vomiting site

There is a classic site on the front of the wrist known as Pericardium 6 (PC6) which has been widely used to treat vomiting. It has been used for post-operative vomiting and for vomiting associated with pregnancy. Good results have been reported and this is probably the classic acupuncture point that has received the greatest amount of research in the West.

The site is over the median nerve. For this reason insertion needs to be shallow, preferably with a 15mm needle. The TCM description states that the site is between the tendons of palmaris longus and flexor carpi radialis, and is three finger-breadths above the distal skin crease. This degree of accuracy is probably unnecessary and not everyone has a palmaris longus.

Wrist bands that apply pressure to this area are sold to prevent travel sickness.

15.5 Site for upper thoracic pain

TCM describes a site on the ulnar border of the hand known as Small Intestine (SI3). It is used for upper thoracic pain and it does appear to work. Presumably this is an example of acupuncture based on body segments since this region of the hand is in the C8/T1 dermatome.

Officially the site is at the neck of the fifth metacarpal and the needle should pass in front of the metacarpal, but it is doubtful that this degree of precision is needed. The treatment can be quite painful, so use a 15mm needle if possible and avoid going in through the palmar skin by pushing it forward (Figure 15.1).

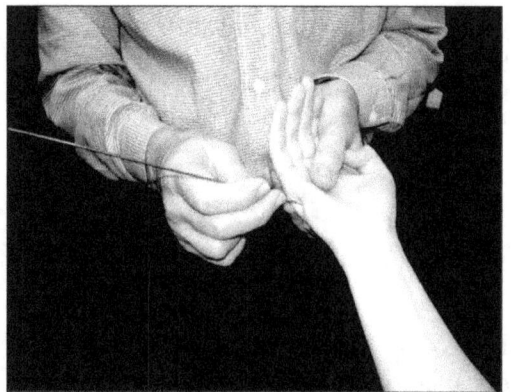

Figure 15.1

15.6   Large Intestine 4 (LI4)

This is probably the best-known classic site in the body. It is situated in the web between the thumb and first finger (adductor pollicis). Both traditionalists and modernists use it frequently as a generalised stimulation site (like LR3) and also to treat various symptoms, including headache, sinusitis, and dental pain. In spite of this I prefer not to use it, for several reasons.

> ▷ Doing so is usually painful.

> ▷ LR3 seems to be more effective.

> ▷ A case of bilateral hand swelling lasting several weeks has been reported.

> ▷ I know of a case of pain in the thumb lasting for six months after this site was needled causing nerve damage (6.6).

> ▷ Two patients in Germany are reported to have lost their hands as a result of using the site. It appears that in these cases the acupuncture was performed too proximally and the radial artery was thrombosed. Normally the radial artery anastomoses with the ulnar artery to form the deep palmar arch, but this is not always the case and then a radial artery thrombosis is a disaster.

For all these reasons I have abandoned the use of LI4. If you do decide to needle it, use superficial needling rather than needling into the muscle.

A convenient way of doing it is to ask the patient to press thumb and first finger together; this brings up a fold of skin on the dorsum of the hand and the needle can be inserted into the cleft so formed.

15.7 Needling the interossei

The interosseous muscles can be needled by inserting needles at about 45 degrees from the dorsal surface. Use 15mm needles if available, since they cause less pain.

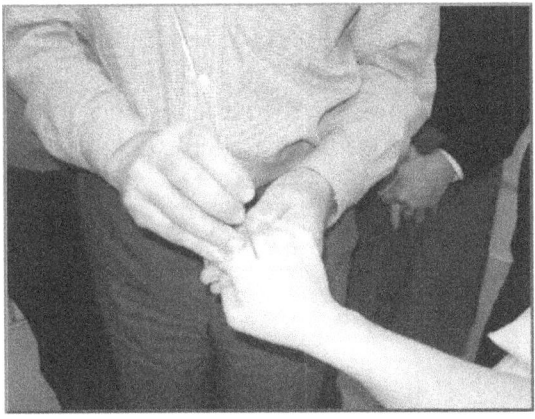

Figure 15.2

This treatment is useful for all kinds of hand pain but especially for osteoarthritis in women after the menopause. Some patients report that the Heberden's nodes associated with this condition become smaller after acupuncture.

Travell and Simons describe a similar injection technique in which the needle is inserted between the heads of the metacarpals. I used this approach initially but it caused a lot of bleeding, so I modified it to needle from the dorsal surface as just described.

15.8 Needling an individual digit

To needle a digit, for example for interphalangeal joint pain, insert the needle midway between the vessels and nerve at the side of the digit and the tendon expansion on top. This means that you insert it at 45 degrees in relation to a cross-section of the digit (Figure 15.3).

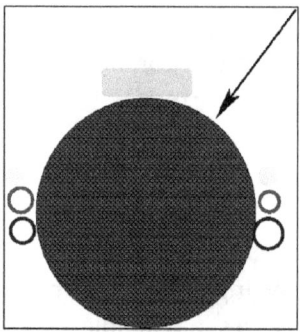

Figure 15.3

Both sides of the shaft, medial and lateral, can be needled. I used to advise using four sites, two on the proximal and two on the middle phalanx, but the distal pair are more painful and I now think that it is enough to do just the proximal sites. The skin should be pulled tightly round the digit to make it taut and to reduce the distance to the periosteum from the skin. Use a 15mm needle if available (10.8).

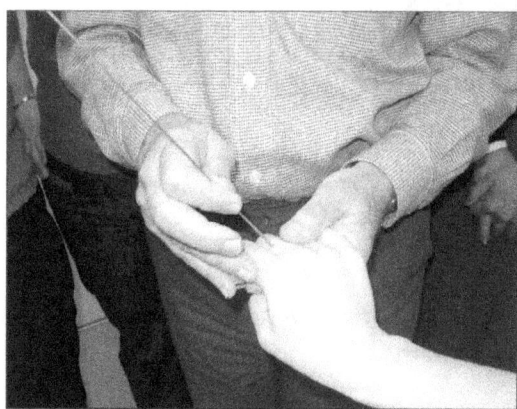

Figure 15.4

15.8.1 Rheumatoid arthritis

Pain due to this disease responds quite well to the treatments described here. But avoid needling round acutely inflamed joints because of the risk of an aggravation.

15.9   Trigger finger

Most patients with this disorder have a nodule at the base of the affected finger. The nodule can be needled with a 15mm needle and the results are generally good.

Chapter 16

Thorax

Chapter Outline

▷ Paraspinal muscles

▷ Sternum

▷ Chest wall

16.1 Introduction

It is often necessary to needle the thorax but care is needed here because of the proximity of the lungs and spinal cord. The main areas to be treated are the paraspinal muscles and the chest wall.

16.2 Needling the paraspinal muscles

The paraspinal muscles can be needled to relieve pain due to ankylosing spondylitis. osteoporosis, and other causes of widespread back pain. In these disorders relief usually lasts about 12 weeks after a few initial treatments.

16.2.1 Technique

The patient may be either sitting or lying prone. Palpate the vertebral spines and place your thumb on each side of the spines. **To avoid pneumothorax don't needle more laterally than this.** Insert needles in line with the middle of your thumb, at about 45 degrees in the sagittal (vertical) plane so as to penetrate the muscles. Once they are all in place, stimulate them briefly once or twice and then withdraw them. Although about 12 needles may be used the treatment is still quite brief.

Traditionalists insert the needles at 'Bladder' points. (The Urinary Bladder channel, the main back channel, runs along each side of the spine in the traditional scheme.) Trigger point enthusiasts will carefully palpate the paraspinal muscles and needle any tender areas they find, but I have been unable to convince myself that doing this makes any difference. I therefore simply needle in a zigzag pattern over the painful area, the needles being inserted about 5cm apart on each side of the spine.

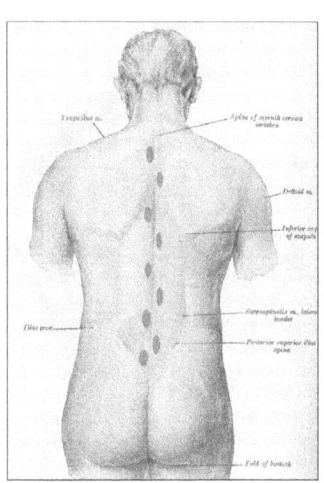

Figure 16.1

When using a fairly large numbers of needles, as in this case, I normally insert them all together rather than each one singly. Occasionally one needle gives rise to continuing pain after it is inserted. If that happens, withdraw it.

To make sure that no needles have been left behind, run your hand lightly over the area after taking them out.

16.2.2   Patients with cancer

Don't needle the back in anyone who you suspect may have **secondary malignant deposits** in the spine. There have been cases of transection of the spinal cord in such circumstances, caused by displacement of a vertebra (spondylolisthesis) due to relaxation of muscles which were splinting the vertebral column.

16.3   Needling the chest wall

There are three safe methods for treating this region: tangentially, lifting up a fold of skin, and periosteally on a rib.

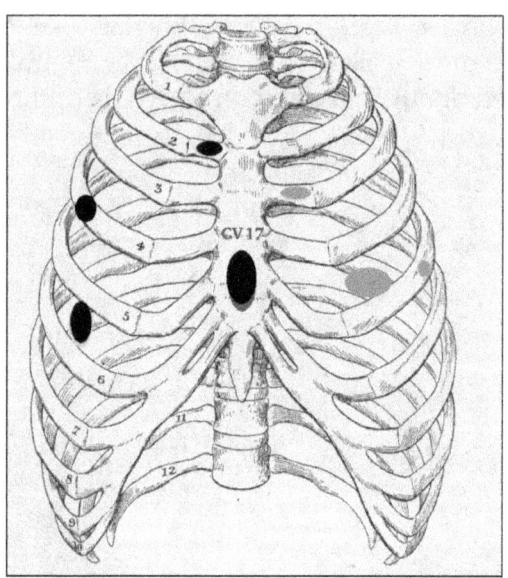

Figure 16.2

16.3.1   Tangential needling

This means that, if the chest is considered as a cylinder, the needle is angled so as to form a tangent to its surface.

16.3.2   Lift up a fold of skin

If the patient is very thin it is safer to pick up a fold of skin before insert the needle tangentially. The disadvantage of this method is that it makes penetration of the skin more difficult and therefore usually more painful, but that is better than risking a pneumothorax.

16.3.3 Periosteal needling on a rib

In a reasonably thin patient it is easy to isolate a rib between two fingers and needle it periosteally, but make sure you can definitely feel the rib and insert the needle **absolutely perpendicularly.** The same technique can he used over a costal cartilage. This can be done if an individual rib or costal cartilage is painful, perhaps after an injury. Use 15mm needles for ribs if available; if the rib cannot be reached with a short needle it is probably safer not to do periosteal needling.

16.4 Chest wall pain

Some patients complain of an area of chest wall which is painful for no apparent reason. The pain arises spontaneously and may persist for many weeks or months.

This disorder responds well simply to superficial (subcutaneous) needling in the area of pain. Three or four needles are inserted and stimulated briefly. **Do not attempt to needle the intercostal muscles**.

16.5 Sternal pain

Non-cardiac sternal pain occurs in some patients and can be treated effectively by needling **subcutaneously** over the sternum. But distinguishing cardiac from non-cardiac pain is important and may be difficult.

As noted previously (6.5), there is a dangerous site (CV17) in the lower sternum where a needle might penetrate the pericardium and cause fatal cardiac tamponade. **Never needle the sternum periosteally.**

16.6 Pectoral muscles

Trigger points may form in pectoralis major. They give rise to chest wall pain which is difficult to distinguish from angina pectoris. Such

patients may be admitted to hospital with suspected heart attacks, but nothing is found to substantiate the diagnosis. Acupuncture can be helpful in such cases. But the situation is complicated because patients who do have angina may also have TPs in their anterior chest wall, and needling these may cause a severe anginal attack requiring urgent admission to hospital. **So be cautious about needling the chest wall in patients who are suspected of having ischaemic heart disease.**

Chapter 17

Abdomen

Chapter Outline

▷ **Needling technique**

▷ **Conditions treated**

▷ **Scar pain**

17.1   Introduction

A number of abdominal disorders may respond to acupuncture. Mostly these affect the lower abdomen, below the umbilicus. I have not had much success in treating upper abdominal disorders such as non-ulcer dyspepsia. Mann reported a similar lack of success with this (26.2.1).

One exception: I have seen two patients with persistent hiccups who were cured by brief superficial needling in the epigastrium.

17.2   Technique

The technique for needling the abdomen is very simple. Insert two lines of needles subcutaneously below the umbilicus. This means that about

8 needles are usually inserted, although there is nothing specific about this. If there is pain in a localised area, such as one lower abdominal quadrant in diverticular disease, insert extra needles there.

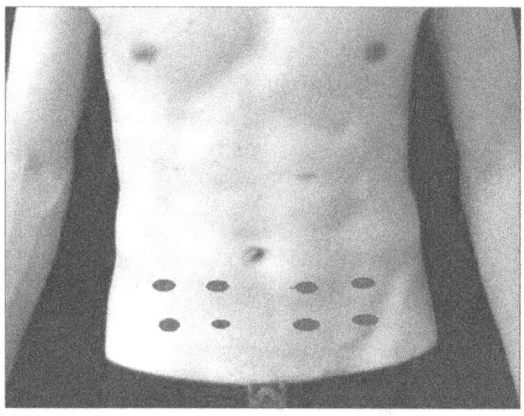

Figure 17.1

In most people there is enough subcutaneous tissue to make horizontal insertion easy to do. If patients are unusually thin a fold of skin can be pinched up and needled horizontally but this is usually more uncomfortable for the patient.

To help you judge the depth of the subcutaneous tissues ask the patient to raise his or her shoulders slightly from the couch so as to contract the abdominal muscles.

17.2.1   Subcutaneous or intramuscular?

Some modernists say that you must needle the abdominal muscles if the treatment is to work. This idea seems to be based on theory rather than practical experience; I have found subcutaneous acupuncture in this region to be fully effective.

17.3   Indications

17.3.1   Ulcerative colitis

This often responds surprisingly well. All the symptoms may diminish or clear up, including pain, bleeding, and diarrhoea. One patient reported that she always had a sudden urgent bowel motion about half an hour after treatment and would then be free of symptoms. She went into complete remission.

Acupuncture is an addition to conventional treatment in this disorder and not a substitute for it, so patients continue to receive standard care. How acupuncture works in this disorder is uncertain. It may be a segmental or humoral effect and a psychological element cannot t be excluded.

17.3.2 Crohn's disease

This is symptomatically quite similar to ulcerative colitis but is pathologically different. There have been a few reports of success with acupuncture but I have treated too few patients to be certain myself.

17.3.3 Diverticular disease

Acupuncture can be helpful for the pain in this disorder. Place extra needles over the site of pain.

17.3.4 Irritable bowel

The classic symptoms of this disorder are pain, bloating, and alternating constipation and diarrhoea. The differential diagnosis includes bowel cancer so you must be certain that this has been excluded before starting acupuncture.

Acupuncture can help with pain and sometimes with chronic diarrhoea. It does not usually do much for bloating or constipation.

17.3.5 Unexplained abdominal pain

Some people have persistent abdominal pain for which no cause can be identified. This is a worrying situation and some of these patients do  eventually turn out to have serious disease which had not been detected. Acupuncture can relieve pain of this kind but the fact that acupuncture is a symptomatic treatment that can mask pathology has to be kept in mind.

17.3.6   Overactive bladder (unstable bladder)

This is a functional disorder characterised by urgency and frequency of micturition. It could be thought of as the bladder equivalent of irritable bowel disorder. Acupuncture is often effective here; use the standard subcutaneous needling technique over the lower abdomen.

Traditionalists use distant points in the lower limb ('Bladder points') to treat these symptoms. I have not found that doing so gives any additional benefit.

17.4   Scar pain

I discuss this topic here for convenience but these remarks apply equally to scars in any part of the body.

The treatment is highly effective. It consists simply in needling the scar itself, particularly at sites of tenderness. A convenient way of doing this is to reverse the needle and use the blunt end as a probe. When you find a site of tenderness, which may be reddened, reverse the needle and insert the point into the scar to a depth of about 3 mm.

As many sites as necessary may be treated in this way. Use 30mm needles, not for the length but for the thickness (to penetrate tough tissues).

The treatment works for both postoperative and post-traumatic scars. It can be done at any time after the scar has healed, which in practice means after about six weeks. It can be effective even in scars that have been painful for many years.

17.4.1   Mechanism of action

This could be thought of as an example of the simplest form of acupuncture, needling the site of pain. But the mechanism of action in this case is probably not local but is due to central changes in the spinal cord or brain. (See Chapter 24 for further discussion.)

Chapter 18

Lumbar and Gluteal Regions

Chapter Outline

▷ **Low back pain without radiation**

▷ **Quadratus lumborum**

▷ **Referred pain to lower limb**

▷ **Gluteal muscles**

▷ **Sacroiliac region**

▷ **Hip pain**

18.1   Introduction

The lower back is one of the most frequently treated areas in acupuncture, with a high success rate. The German GERAC trial published in 2007 found that the effectiveness of acupuncture for chronic low back pain was almost twice that of conventional therapy. It was also surprisingly long-lasting—at least six months. But there was no difference between 'real' acupuncture and acupuncture done superficially at non-acupuncture sites (25.2.1).

18.2   Low back pain without radiation

Pain confined to the lumbar region can be treated either by needling the paraspinal muscles or by inserting a row of 30mm needles (usually four) at the L4 or L5 level. There is no need to identify individual trigger points.

Since the spinal cord terminates at about the level of the lower border of L1 and the muscles in this region are large, there are no special anatomical safety considerations for this treatment.

18.3   Quadratus lumborum

This is an important muscle to treat in people who are active physically, such as those sports people and ballet dancers.

18.3.1   Technique

The patient lies on the side with the affected side uppermost. Place a pillow underneath the spine to flex it laterally and open up the space between the lower ribs and the iliac crest.

The hip nearer to the couch is flexed and the uppermost limb is more or less straight. Quadratus lumborum can then be palpated. If there is any doubt about its location ask the patient to raise the shoulders off the couch so as to tense the muscle.

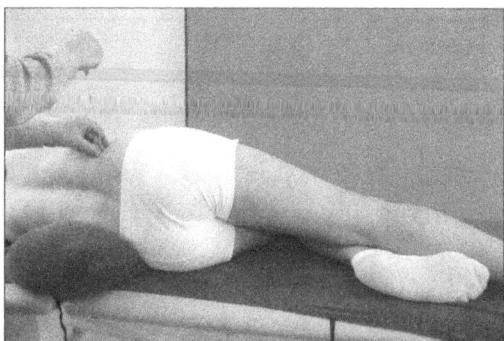

Figure 18.1

The needle should be directed towards the lower spine at about L4 level. **It should not be directed upwards because of the risk of injuring the**

kidney. To make sure of the correct direction, palpate the lumbar spine with one hand and use this as a landmark. Usually a 30mm needle is long enough although in large patients you may need a 50mm needle.

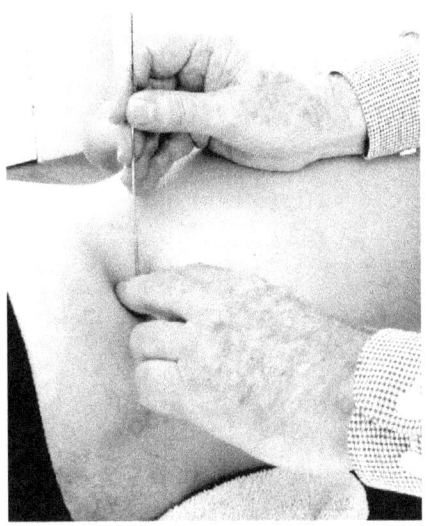

Figure 18.2

18.4 Referred pain to lower limb—'sciatica'

The term 'sciatica' should mean pain in the distribution of the sciatic nerve although it is often used more loosely to describe any referred pain in the lower limb.

18.4.1 Pathology

The cause of referred pain in the lower limb is often uncertain in spite of frequent confident statements to the contrary. In some cases there is pressure on a nerve root, due for example to lumbar spondylosis or a prolapsed intervertebral disk, but that is not always the case. In at least some patients the problem seems to be in the gluteal muscles such as piriformis (see below).

There may be accompanying sensory or motor changes, including anaesthesia and muscle weakness. Acupuncture is unlikely to help if these are present although it is still be worth trying because there is nothing to lose and sometimes the results confound expectations.

18.4.2 Treatment options

There are two main forms of treatment in this region: needling the gluteal muscles and needling the periosteum in the sacroiliac region.

18.5 Gluteal muscle needling

The most commonly cited muscle here is piriformis although others, such as gluteus medius or gluteus minimus, may also figure.

18.5.1 Piriformis syndrome

The so-called piriformis syndrome is supposed to be due to entrapment of the sciatic nervee by piriformis. Usually the nerve passes deep to piriformis but in some people it runs partially or wholly through the muscle. Contraction of piriformis is then supposed to compress the nerve and give rise to symptoms. The reality of this mechanism is uncertain, and the symptoms could equally well be due to radiation from a TP in piriformis or other gluteal muscles. Whatever the exact explanation may be, needling the gluteal muscles is often effective.

The following description applies to needling any of the muscles in this region: piriformis, gluteus medius, and gluteus minimus.

18.5.2 Examination technique

The patient lies on one side with the affected limb uppermost and flexed; the opposite limb is more or less straight.

> ▷ Particularly for women, place a pillow under the knee to prevent excessive rotation.

> ▷ Adjust the position of the affected limb so as to place the gluteal muscles under slight but not excessive tension.

> ▷ Palpate the muscles, gently at first but then with increasingly firm pressure, culminating in the use of both hands to exert firm compression of the tissues.

▷ This will elicit tenderness and possibly radiation of sensations (not always pain) down the limb and will allow you to feel bands of contracted muscle.

▷ If the patient reports that this palpation causes sensations in the areas where the sciatic pain is felt this indicates that the treatment is likely to succeed, although it may work even without this.

▷ If TPs are found they should be visualised. They probably range in size between a hazelnut and a walnut.

18.5.3 Needling technique

▷ Compress the overlying tissues with your free hand.

▷ Insert a needle (usually 50mm length) into the target.

▷ When this is reached the patient will typically experience a particular sensation: sometimes pain, more often a deep feeling of pressure.

▷ You may feel a change in resistance accompanied by a curious 'grating' or 'creaking' sensation.

▷ If this doesn't happen you can partially withdraw the needle, alter its angle, and try again. But don't prolong this process too much.

▷ The patient may experience radiation down the limb which may correspond to the distribution of pain. This is a favourable sign.

▷ Depending on how much effect you are producing you may now increase the angle of twist until you feel the needle being firmly gripped. It feels as if the fibres are being wound up round the needle. At the same time the patient will feel an increase in sensations.

▷ At this point I sometimes say to the patient "Please bear with it for a moment". After about 10-20 seconds the muscle begins to relax and the patient's sensations start to diminish. I then remove the needle.

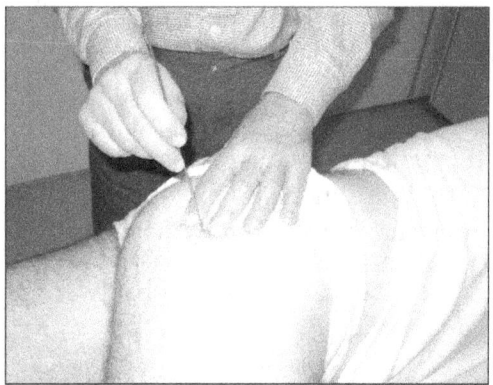

Figure 18.3

18.5.4   Trapped needle

Rarely the needle may be locked after this and refuse to come out. In that case leave it alone for a few minutes, after which the muscle relaxes and the needle can be withdrawn, perhaps after a few gentle twists to free it (10.13.1). Traditional books often advocate inserting another needle close by to relax the muscle, but I prefer to use the method described here.

18.5.5   Treating multiple sites

Sometimes there is more than one TP but it is probably a bad idea to treat more than one site, at least on the first occasion.

If patients are extremely sensitive even to light touch (this is unusual), use superficial needling on the first occasion instead of the technique just described.

18.5.6   Traditional terminology

 The site known as Gall Bladder 30 (GB30) overlies the piriformis. It could be used as a shorthand description of where to needle but I don't do so, for several reasons.

> ▷ The site of maximum tenderness may not be over piriformis but over a different gluteal muscle.

▷ GB30 refers to a surface location and what we are interested in is targets at a considerable depth below the surface.

▷ Using this name implies an unwarranted degree of precision. In reality the potential target is considerably larger than this.

Nevertheless, GB30 is significant in another way. It is described as located one-third of the distance from the greater trochanter of the femur and the sacral hiatus (the notch at the bottom of the sacrum where the cauda equina comes out). Below this line you may hit the sciatic nerve if you needle too deeply; above the line the nerve should be safe. (In most cases accidentally needling the sciatic nerve will cause only temporary symptoms.)

18.6 Periosteal sacroiliac needling

This treatment consists in needling the periosteum in the region of the sacroiliac joint. (The needle doesn't penetrate the actual joint but it is convenient to use the abbreviation SIJ for the treatment.) Depending on the depth that is reached the needle may penetrate the sacroiliac ligament and reach the periosteum.

18.6.1 Technique

The surface landmark is the gap between the sacrum and the ilium, at or below the level of the posterior superior iliac spine.

Felix Mann, who first described the technique, had the patient sitting and leaning forwards on a support. This is how I usually do it myself but it can also be done with the patient prone or side lying (if only one side is to be treated).

18.6.2 Locating the needling site

Place your fingers on the iliac crest and use your thumb to identify the gap between the ilium and the sacrum. This is more medial than you might expect and if you have a small hand you may need to move it to find the correct site of insertion.

18.6.3   Needle insertion

The exact technique will vary according to the patient's position (prone, side lying, or sitting and leaning forwards).

If the patient is sitting and leaning forwards the angle of the needle must be adjusted to allow for this; usually that means that you will have to aim it slightly downwards, otherwise it will be pointing cranially so far as the patient is concerned.

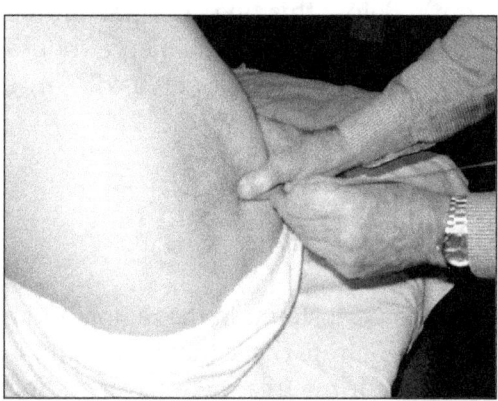

Figure 18.4

The needle is advanced until it reaches the periosteum, either on the back of the sacrum or on the side of the ilium. Since this is not being done with imaging control we cannot be certain exactly where it will reach the periosteum; a slight variation at the surface will make a considerable difference at depth, and the needle itself is flexible. Nevertheless the treatment is safe since there are no critical structures in the region.

There are wide variations in the depth of the pelvis at this point. In some patients you will reach the periosteum quite quickly, in which case a 30mm needle might be long enough; in others the periosteum will not be reached at all even with a 50mm needle.

Use a guide tube?

For beginners it is easier to insert the 50mm needles with a guide tube to prevent excessive flexing (10.16).

18.6.4   Patient sensations

There are surprisingly wide variations in what patients feel in response to this treatment. A few find it painful but this is unusual. Many feel nothing or almost nothing. Others feel the characteristic periosteal sensation of deep aching. There is often radiation which may go down the limb in various ways.

18.6.5   Unusual pain patterns

Although disorders of the sacroiliac and gluteal regions usually cause radiation to various levels in the lower limb, they occasionally cause pain in either the scrotum or or a lower abdominal quadrant. Pain in these areas can sometimes be treated effectively by needling the gluteal region or SIJ but be sure that you have excluded serious pathology, which may be difficult.

18.7   Choice of treatment

As already mentioned, the sacroiliac and gluteal sites can be combined but often the clinical picture and the results of physical examination will suggest which method is more likely to be successful. The presence of active TPs in the gluteal muscles would suggest starting there while their absence might favour the sacroiliac approach. On the first occasion it is usually preferable to use one method only until the degree of response becomes apparent. (Remember the golden rule: do less rather than more.)

18.8   Repetition of treatment

Sciatica is a disorder in which you might give a number of treatments, perhaps as many as 12, provided there is at least some response after the first two or three sessions. Many patients may require occasional 'top-up' treatments to maintain improvement, although one session is usually enough on these occasions to bring about a remission once more.

18.9   Needling an individual vertebra

Occasionally you may wish to needle a vertebra. This is possible below the level of the spinal cord termination (usually the lower border of L1). To do this, identify the vertebral spine and insert a needle slightly lateral to it, aiming it medially so as to reach the side of the spine or possibly the lamina. **Try to avoid needling the facet joints**; there has been a case of septic arthritis of such a joint caused by acupuncture.

18.10   Hip joint

Osteoarthritis of the hip can often be helped symptomatically for a time. Needle the greater trochanter of the femur periosteally, with the patient in the same position as described for needling the gluteal muscles. It can be reached with a 30mm needle in a thin patient but it is surprising how often a 50mm needle is required here, especially in women.

Chapter 19

Knee

Chapter Outline

 ▷ **Osteoarthritis**

 ▷ **Iliotibial tract syndrome**

 ▷ **Other knee sites**

19.1   Introduction

There are several conditions in and around the knee that respond well to acupuncture. There is clinical research evidence for its efficacy (25.2.2, 26.4.1) in knee osteoarthritis, which is common in middle-aged and elderly patients.

19.2   Clinical features of knee osteoarthritis

 ▷ There may be fairly short-lasting morning stiffness (not more than 30 minutes in duration).

 ▷ There is no palpable warmth and only a small effusion if any.

▷ There may be bony enlargement, bony tenderness, and muscle wasting.

▷ The differential diagnosis includes gout and pseudogout, haemarthrosis, avascular necrosis, and malignancy. In these conditions there is likely to be more effusion and warmth. In the case of malignancy the pain is likely to be worse at night.

▷ Don't forget the well-known trap of referred pain from the hip or lower back.

▷ X-ray changes may not always correlate well with the clinical findings

19.3   Acupuncture for knee osteoarthritis

▷ Since this is an an intrinsic knee disorder the principles of acupuncture suggest that periosteal needling should be used. The most accessible site is the flat area at the medial side of the tibia below the knee.

▷ This area has no conventional anatomical name; Mann called it the infragenual area.

▷ There may be soft-tissue tenderness, particularly in middle-aged women, but you don't have to needle the tender area specifically because this is a periosteal technique for which local tenderness is not a guide to treatment.

▷ In most people a 30mm needle is adequate but in some cases there is thick overlying tissue and then a 50mm needle may be required.

▷ Peck the periosteum firmly but quickly two or three times.

▷ Some patients find this quite painful while others hardly feel it at all. But in any case the treatment is very quick, lasting only a few seconds.

▷ Usually up to six treatments are required, initially at weekly intervals. There is then a period of remission lasting 8 to 12 weeks, sometimes longer.

▷ Patients with this type of knee pain should always be taught quadriceps-strengthening exercises.

19.3.1   Alternative: needling the patella

You can also needle the anterior surface of the patella periosteally. This may work for patients whose pain seems to be localised behind the patella. It does not work for the usual kind of osteoarthritic knee pain, probably because the patella is not really part of the knee joint but is a large sesamoid bone in the patellar ligament.

Needling the patella is often more painful than needling the tibia, although the reverse may also occur.

19.4   Iliotibial band (tract)

This structure is a continuation of the fascia covering tensor fasciae latae. It crosses the knee joint and helps to stabilise the knee in extension. In runners, walkers, and cyclists it can give rise to pain when their activity is increased beyond the habitual level and it is subjected to friction.

Typically the patient feels pain over the knee after exercise. When you examine the knee you will probably find a tender area just where the band crosses the knee joint. The landmark is the fibular head; palpate above and in front of this to find the band. If in doubt ask the patient to abduct the foot to tauten the band.

19.4.1   Needling technique

Insert a needle into the band at the site of maximum tenderness and stimulate it manually for about 30 seconds. A 30mm needle should be used, not for the length but for the thickness. **Since the site of tenderness is directly over the joint the needle should not penetrate more than about 3 mm.** This is an effective treatment.

19.5   Lateral thigh

In some people TPs develop in the lateral side of the thigh in vastus lateralis which may radiate to the knee. These are safe to needle although manual pressure may be more effective. Superficial pain on the lateral side of the knee may be due to a TP in tensor fasciae latae where it arises from the anterior part of the outer lip of the iliac crest.

19.6   Medial thigh

There is always a latent TP in the medial side of the thigh (vastus medialis), about a hand's breadth above the medial condyle. It can give rise to knee pain in young patients ('teens or very early twenties). Needling this site is often effective in this age group although not in older patients.

19.7   Semimembranosus TP

Patients such as typists who sit for long periods on a hard chair which presses on the back of the thigh may activate a TP in semimembranosus. This has an unusual pain referral pattern, causing radiation to the front of the thigh above the patella.

19.7.1   Identifying semimembranosus

▷ The trigger point forms in semimembranosus just lateral to the tendon of semitendinosus.

▷ Semimembranosus is in front of semitendinosus, which you can feel with your thumb.

▷ Semitendinosus runs in a groove on the back of semimembranosus, which is therefore palpable lateral and anterior to semitendinosus.

▷ The patient should lie on the side with the affected knee uppermost and a pillow between the knees.

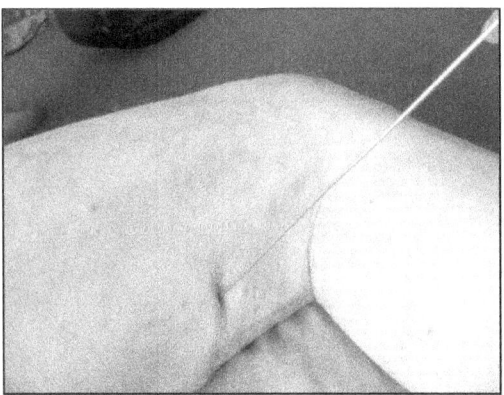

Figure 19.1

19.7.2 Needling technique

Insert a 30mm needle obliquely into semimembranosus in front of semitendinosus. The direction is forwards and medially, therefore away from the popliteal vessels and nerve.

> The description of this site can be confusing. The TP is on the *medial* side of the thigh but *lateral* to semitendinosus. The needle points *antero-medially*, which is *away from* the midline of the thigh.

19.8 Traditional knee treatment

The traditional treatment sites called the 'eyes of the knee' are found on either side of the patellar ligament. Although they are widely used, they entail the risk of penetrating the knee joint and thus causing a septic arthritis. I therefore think that these sites should not be needled, especially since effective alternatives exist.

Chapter 20

Leg

Chapter Outline

▷ **Needling technique**

▷ **Leg pain**

▷ **Plantar fasciitis**

▷ **Achilles tendinopathy**

20.1 Introduction

In the anatomical sense, 'leg' refers to the region below the knee as far as the ankle. Good results are obtained by needling here in a number of disorders.

20.2 Needling technique

Insert several 30mm needles into the calf muscles (soleus and gastrocnemius) on both sides of the midline. I usually place three on each side.

If there are any TPs these may be needled selectively but usually this is unnecessary.

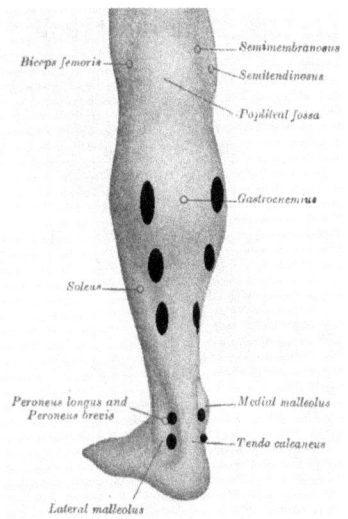

Figure 20.1

20.3 Conditions treated

20.3.1 Leg pain of various kinds

Leg pain of any kind, especially ischaemic leg pain (including intermittent claudication), responds well.

20.3.2 Diabetic leg pain

This responds in about half the cases, possibly more. **Be particularly careful about your hand washing in this case owing to the increased risk of infection.**

20.3.3 Restless leg syndrome

This often responds to the same treatment as that used for pain.

20.3.4   Plantar fasciitis

In some patients this condition is due to trigger points in the gastrocne-mius or soleus and can be treated by needling those. But most require direct needling of the sole (21.7).

20.3.5   Shin splints

These are reported to respond to periosteal needling along the length of the tibia.

20.3.6   Stump pain

Patients with pain in the stump after amputation may respond to su-perficial needling of the healed stump.

20.3.7   Phantom limb pain

Needle the opposite limb in the area corresponding to where the pain is felt in the phantom. This can be remarkably effective and has been associated with disappearance of the phantom in at least one case. Such effects are evidence for a central response to acupuncture (26.3.8).

20.3.8   Achilles tendinopathy

This condition responds quite well to acupuncture. In most cases it affects the middle part of the tendon.

   ▷ Insert 2 or 3 small (15mm length) needles on each side of the tendon, angling them inwards to form a V. **Don't needle the tendon itself since if done repeatedly it will break the fibres and may increase the risk of tendon rupture**.

   ▷ If there is an apparent swelling in the middle of the tendon, sur-round it with the needles. Acupuncture doesn't seem to reduce the swelling much in these cases but it does relieve the pain.

▷ Occasionally the problem is not in the middle of the tendon but is situated either proximally, at its origin, or distally, at its attachment to the calcaneus. In those cases the needling should be done at the affected area.

20.4   Classic acupuncture points in the leg

The treatments just described do not correspond to classic acupuncture points. But there are some points in this region which you should know about because they are widely used and spoken about, even by some modernists.

▷ **Spleen (SP6)** is on the medial side of the leg about a hand's breadth above the medial malleolus. In the traditional system it is supposed to be related to the female pelvic organs. It is used to treat primary dysmenorrhoea (painful menstruation in young girls) and is usually effective for this. But this is not evidence for point specificity, since it is not clear whether the same effect would be produced by needling, say, LR3 or the lower abdomen.

In TCM SP6 is one of the 'forbidden areas' in pregnancy because it is supposed to be capable of inducing abortion. There is no good evidence that this is a real risk (and some evidence that it is not).

▷ **Bladder 57 (BL57)** is in the midline of the leg, in the angle between the two parts of the gastrocnemius. I have not found that using this makes a difference to the results when needling the calf muscles.

▷ **Gall Bladder 34 (GB34)** is situated close to the neck of the fibula and the superficial peroneal nerve.

▷ **Stomach 36 (ST36)** is anterior and inferior to GB34, in peroneus longus.

▷ Both the last two sites, especially GB34, are liable to cause nerve damage, which is a good reason to avoid them. A case of foot drop has been reported. Nevertheless they are widely used by TCM practitioners and by modernists who believe in traditional points.

ST36 and GB34 are both often cited as having dramatic effects in various disorders, including shoulder pain. I think the probable explanation is that in a strong reactor almost any site in the leg may have pronounced central effects (12.2.1).

Chapter 21

Ankle and Foot

Chapter Outline

▷ **Periosteal needling**

▷ **Hallux pain**

▷ **Interdigital neuroma (Morton's neuroma)**

▷ **Plantar fasciitis**

▷ **Plantar warts**

21.1 Introduction

The region of the ankle and foot is rewarding to treat with acupuncture. There is a high success rate, including in some conditions for which there is little effective conventional treatment.

21.2 Non-specific instep pain

Many middle-aged and elderly patients complain of pain in the instep region which is often described as osteoarthritic although radiological evidence is usually absent. This pain responds well to acupuncture.

139

21.2.1   Technique

You can needle the periosteum on any of the foot bones wherever it is accessible and safe to do so. There should be no risk of penetrating a joint and no vulnerable structures to be injured. Good targets include the navicular, calcaneus, cuboid, and medial and lateral cuneiform.

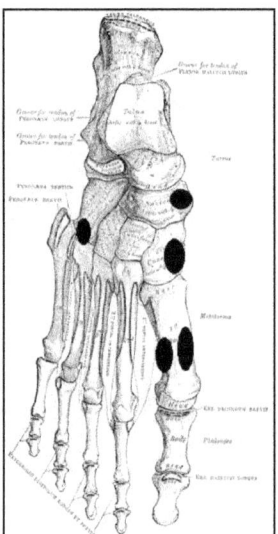

Figure 21.1

21.3   Hallux pain

This responds well to periosteal needling to the shaft of the first metatarsal. The technique is similar to that used for the fingers (15.8). Insert a needle on each side of the metatarsal, between the dorsal tendon expansion and the vessels and nerve.

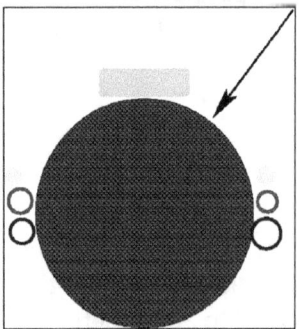

Figure 21.2

It is surprisingly easy to miss the shaft of the metatarsal while doing this treatment, especially if the patient is much overweight. Identify the bone distinctly before needling it. You can needle the first phalanx in the same way as the metatarsal, but this is probably unnecessary and is more painful.

After one or two sessions most patients get a remission lasting about 12 weeks so treatment is usually required about four times a year.

The same technique can be used for other toes if necessary.

21.4 Sprains

Sprains, both acute and chronic, can be treated effectively by periosteal needling at the attachments of the ligaments (usually the anterior talofibular and calcaneofibular ligaments). This is a symptomatic effect and you should decide on clinical grounds what activity restrictions, if any, to advise.

21.5 Interdigital neuroma (Morton's neuroma)

This often responds well to acupuncture. The treatment consists in inserting a 30mm needle from the dorsum of the foot into the neuroma and stimulating it manually.

21.6 Plantar fasciitis

This painful disorder is generally difficult to treat by other means but often responds very well to acupuncture. Some patients respond to needling of trigger zones in the calf muscles (20.3.4), but most require direct needling of the attachment of the plantar fascia to the calcaneus, usually at the medial calcaneal tubercle.

21.6.1 Technique

The aim is to reach the periosteum of the calcaneus. There are two ways to do this, direct and oblique.

21.6.1.1   Direct route

In this method you go vertically through the heel skin until you reach the periosteum. You then peck it in the usual way.

This can be difficult to do because of the thickness of the plantar skin and the denseness of the underlying tissues. Some podiatrists have therefore modified the technique to use the oblique route.

21.6.1.2   Oblique route

To use this method start a little distal to the tubercle, where the skin is thinner and therefore easier to penetrate. Angle the needle at about 45 degrees, aiming towards the periosteum. Don't go too medially or you may miss the calcaneus.

21.6.1.3   Choice of needle for plantar fasciitis

In most cases a 30mm needle is long enough to reach the periosteum although some patients with large feet may need a 50mm needle. Use the shorter needle if possible since it is easier to manipulate.

21.6.1.4   Acceptability to patients

The treatment is moderately painful as a rule, but patients are generally in such pain already from their condition that they are willing to have the treatment anyway. Usually a complete and permanent cure is obtained after 1 to 3 treatments although a few people do require repeats later.

21.7   Plantar warts (verrucas)

A number of podiatrists who use acupuncture have tried needling plantar warts and in some cases have reported dramatic cures of warts that had been present for months or even years. It seems unlikely that this is an acupuncture effect of the usual kind. Probably it works, if it does, by carrying virus particles into the blood stream and eliciting an immune response.

Chapter 22

Recording the Treatment

Chapter Outline

▷ **What needs to be recorded**

▷ **The first treatment**

▷ **Subsequent treatments**

▷ **Examples**

22.1   Introduction

It is clearly vital to make adequate clinical records, both for your own purposes and for anyone else in your clinic who may treat the same patient in the future.

Another reason for keeping good records is medicolegal. If a patient makes a complaint your notes will be essential for answering the complaint and their quality may be decisive in the final outcome.

In what follows I provide some guidelines to what to record. You will have your own method of recording your cases and I merely cover the specific aspects that relate to acupuncture. What I describe here is the bare minimum you need; you can and probably should record more than this in most cases.

22.2 *At the first visit*

22.2.1   History

This is much the same as your standard history except that if the patient
has had acupuncture previously you want to find out the details of
what was done, how it was done, and what the outcome was.

22.2.2   Presentation (diagnosis/symptoms)

Always try to make a pathological diagnosis if possible but don't go
beyond what can reasonably be inferred from the presenting features.
Quite often it is more accurate and honest simply to record the patient's
symptoms.

For example, it is meaningless to put 'cervical spondylosis' for a patient
who comes with neck pain.without radiation. But if a patient is suffering
from definite rheumatoid arthritis, that is a valid diagnosis.

22.2.3   Medication

Get a full list of what the patient is taking, including any over-the-
counter preparations. Don't forget herbal medicines and the like.

22.2.4   Examination

Your physical examination should be particularly directed to range of
movement and tenderness. Remember also to look at the patient's skin.
Many skin disorders also involve the joints joints (e.g. psoriasis) and
you should be careful about needling through potentially infected skin
(which includes eczema and psoriasis).

22.2.5   Pretreatment checks

Note any red flags, contraindications, or special precautions that are
indicated in this case. Also make a note of the patient's consent to
treatment.

22.2.5.1 Recording consent

If you don't get written consent make sure you have obtained verbal consent and record the fact in writing. Remember that the validity of consent depends on the adequacy of the information you give and not on whether consent was written or verbal.

If the patient asks you any questions about safety, write down verbatim what you were asked and what you answered. This will be important in the event of a subsequent complaint.

Give the patient a written information sheet.

22.2.6 Recording needle insertion sites

Traditionalists can presumably write down a list of points treated to record where the needles have been placed. Non-traditionalists don't have this option so what is the solution?

I do use a very small number of 'points' as shorthand to say where needles have been inserted: GB20, GB21, SP6, LR3 (1.7). You may use these or different ones or none at all, but in any case if you are not a traditionalist you will need some other way to describe what you have done. There are two possibilities:

▷ Verbal description

▷ Marking the sites on a drawing

I mostly rely on verbal description. Outline body drawings which don't show muscles and are not precise enough to be useful.

22.2.7 Depth of insertion

It doesn't make sense to specify the depth in millimetres. But depth should be recorded if it is unusual (either periosteal or superficial).

22.2.8 Needle sizes

Most needling is done with 30mm needles so this need not be recorded every time. But if either 50mm or 15mm needles are used this should be noted.

22.2.9   Immediate effects

Note any immediate effects, good or bad. This will include pain, immediate improvements, strong reactor effects, and anything else that seems significant.

22.2.10   Continuation plan

Include here any other comments you want to make.

22.3   *When the patient returns*

It is important to get a narrative account of what happened after the last treatment. There is a tendency for patients to give simple statements (better, worse, the same), but don't be satisfied with this. Many patients report that there has been no effect although on questioning they say that the symptoms remitted completely for a day and then returned. They regard this as a failure but it is really a good response to an initial treatment.

It is useful although not essential to use some form of pain quantification such as the visual analogue (VAS) scale.

22.3.1   Treatment this time

Often you will repeat the initial treatment unchanged, in which case you can simply note this. Otherwise,record any changes you make.

22.3.2   Shorthand keys

I use the following abbreviations.

L = left
R = right
PO = periosteal
VAS = visual analogue scale
SIJ = sacroiliac joint region

Intensity of treatment effects is roughly indicated on a 'plus scale' (+, ++, +++, ++++).

22.4 Examples

Patient 1: Sciatica

Presentation

Backache (now improved).
Persisting radiation of pain down L leg for 3 months.

On examination

Lumbar spasm ++.
Straight leg raising 80 degrees R, 50 degrees L.
Reflexes and sensation intact.
Tenderness +++ right gluteal region (piriformis/gluteus medius).
VAS = 8/10

Pretreatment checks

Red flag features excluded.
No contraindications to acupuncture.
Patient's consent to acupuncture obtained.

First treatment

SIJ (50mm needle, PO).
Gluteal TP in L piriformis/gluteus medius (50mm needle).

Immediate effects

Radiation of sensations in leg correspond to pain distribution (favourable sign).

Continuation plan

See after one week.

Second visit

Treatment outcome

Dramatically better after last treatment for two days; now relapsing.
VAS = 5/10.

Treatment this time

Repeat treatment as previously. (Note: no increase in the number of needles or the intensity of treatment.)

Continuation plan

See after one week.

Third visit

Treatment outcome

Much better.
No relapse this time.
VAS = 2/10

Treatment this time

Repeat treatment as previously.

Continuation plan

See after two weeks.

Fourth visit

Treatment outcome

Almost no pain now.

Continuation plan

No treatment this time. Discharge.

22.5 *Patient 2: Acute neck pain*

Presentation

Severe pain R side of neck for 10 days.
No radiation to arm or head.
VAS = 7/10

Pretreatment check

No red flags.
No contraindications.
Verbal consent obtained.

On examination

Neck rotation R = 20 degrees, L = 60 degrees.
TP +++ right side of posterolateral neck (semispinalis capitis).

First treatment

GB21 L+R.

TP R side of neck (thin patient—15mm needle).

Immediate effects

Patient feels faint!
Rotation R immediately improved to 30 degrees.

Continuation plan

See after 1 week.

Second visit

Treatment outcome

Initial aggravation for 24 hours, with pain up to 9/10. Then better than before treatment.

On examination

TP R side of neck still present but less active. So last treatment was correct but too strong.

Treatment

NB. Reduce treatment because of aggravation and feeling faint.
GB21 R only.
TP R side of neck.
Both lightly, 2-3 seconds, 15mm needle.

Immediate effect

Slightly dizzy but no fainting this time.

Continuation plan

See after 1 week.

Third visit

Treatment outcome

Much better.
Only slight aggravation for 2 hours (acceptable).
Rotation 60 degrees L+R.
VAS = 1/10.

Continuation plan

No treatment this time. Discharge.
NB. Very mild symptoms at this stage will probably resolve spontaneously over the next few days.

Patient 3: Plantar fasciitis

Presentation

Pain in sole of foot—5 months.
Worst first thing in the morning after getting out of bed.
VAS = 10/10.

On examination

Tenderness +++ over L calcaneal tubercle.
Several TPs L calf muscles.

Pretreatment check

No red flags.
Patient apprehensive about acupuncture but agrees to have it after discussion.

First treatment

Avoid needling sole in view of apprehension.
Needle TPs in L gastrocnemius/soleus.

Continuation plan

See after 1 week.

Second visit

Treatment outcome

No change after last treatment.
Explain situation; patient now willing to try acupuncture to sole.

Treatment

Needle L calcaneal tubercle PO. Use 30mm needle inserted obliquely.

Immediate effect

Minor discomfort only.

Continuation plan

See after 1 week

Third visit

Treatment outcome.

Dramatically better for 2 days, now pain returned but VAS down to 4/10.

Treatment

Repeat acupuncture to calcaneal tubercle.

Continuation plan

See after 2 weeks.

Fourth visit

Treatment outcome

Pain-free since last treatment.

Continuation plan

Discharge.

Chapter 23

Self-Acupuncture

Chapter Outline

▷ **Possible reasons for self-acupuncture**

▷ **Possible objections**

▷ **Practical aspects**

▷ **Patient instructions**

23.1 Introduction

There are two main reasons why you might think of teaching a patient or a patient's relative to carry out the treatment.

23.1.1 Short duration of effect

This is probably the commonest reason for considering self-acupuncture. Some patients respond well but the improvement doesn't last. Each patient seems to have his or her response pattern, and in some cases it becomes apparent that the benefit does not last long enough to make it worth while continuing. As a rule of thumb in a hospital setting I tended to regard 8 weeks as the cutoff point, after which I would begin to think of teaching the patients to do the treatment for themselves.

23.1.2   Difficulty in coming for treatment

Even when the improvement after each treatment lasts more than 8 weeks there may still be problems. Perhaps the patient lives a long way off, or perhaps there are difficulties in paying for the treatment either on the NHS or privately.

23.2   Possible objections to self-acupuncture

23.2.1   Too dangerous?

Some practitioners are worried about the potential medicolegal aspects of teaching patients to treat themselves. This is understandable but I think the risk is not really very great. Diabetic patients are taught to inject themselves with insulin two or more times daily so it is difficult to see why a sensible patient should not put a 15mm needle into their foot perhaps once a week.

23.2.2   Need for frequent treatment revision?

Another objection that is sometimes raised is that the treatment needs to be revised at each visit. This objection usually comes from traditionalists, for whom acupuncture is a complicated esoteric treatment. Since I don't believe this I am not concerned. Once the best way of managing a patient has been arrived at, there normally no further change thereafter and the treatment is the same each time. But patients are told to ask for a further appointment if the acupuncture ceases to work.

23.2.3   Conclusion

A number of hospitals and GPs have taught their patients to give their own treatment, with good results and no complications. I did so myself for 20 years when I was in hospital practice. I therefore think it is safe to do this but I realise that not everyone will be happy with the thought and I would not say that anyone who feels like this should try to overcome their reluctance.

Of course, you continue to be responsible for the patient so far as acupuncture is concerned.

23.3   Practical aspects

When you are thinking about self-acupuncture for a patient you need to consider two things:

> ▷ Is the condition suitable for acupuncture?

> ▷ Is the patient (or a relative) willing and able to do it?

23.3.1   Suitable condition

The patient must already have found that acupuncture works for him or her at least for a time after each treatment. And the treatment that is to be used must be anatomically safe, so needling over the chest wall would not be suitable for self-acupuncture. Many patients only need to use LR3, which is pretty safe.

23.3.2   Suitable patients

Patients must be willing to do the treatment themselves or have it done by a relative. The person who is going to do the acupuncture (usually the patient) must be capable of understanding and following the instructions.

23.4   Suggesting self-acupuncture to the patient

Although a few patients reject the idea of treating themselves most accept it enthusiastically. By this time they have already had a number of treatments in the clinic and know that acupuncture is quick and easy to carry out.

23.5   The training session

Bring the patient (or patient plus relative) in for a slightly longer session. Do the treatment as usual and then ask the patient to do it as well. If the site to be needled is bilateral, as it often is (LR3), you do one side and they do the other.

23.6 Beginning at home

Provided this goes well, as it nearly always does, the patient can start using the treatment at home. You should explain which needles to get and where to get them; my hospital used to sell the needles. In almost all cases patients should use 15mm needles, which reduces the chance that anything will go wrong.

Patients should be able to reach you by email or telephone in case of any queries.

23.7 Needle disposal

There needs to be some arrangement in place for disposal of the used needles. Probably the local council has a scheme of this kind; alternatively, patients may have to pay for collection and disposal but this will not be needed very frequently since only small numbers of needles will be used.

23.8 Follow-up

The patient comes back for review in about a month's time. If all is going well a further appointment should come after perhaps a further 3 months, and the patient can continue to be seen perhaps two or four times a year. If this is difficult he or she can maintain contact by telephone or email.

23.9 Success rates

Self-acupuncture usually works well although some patients say they want an occasional 'top-up'. Adverse effects, in my experience, are exceedingly rare.

23.10   Instruction sheet

It is essential to give patients written information about how to do the treatment and what complications to look out for. Here is an example.

INSTRUCTIONS FOR SELF-ACUPUNCTURE

How to do it

1. Wash your hands in the usual way.

2. Remove the needle from its envelope without bending it. (If you do bend it, discard it and use a fresh one.)

3. Don't use a needle that has been dropped on the floor.

4. Hold the needle in your right hand (if you are right-handed).

5. Stretch the skin with your other hand.

6. Rest the tip of the needle against the skin.

7. Press the needle *quickly* right through the skin without hesitating.

8. If necessary push the needle more until it has gone in about half an inch (10 mm) but avoid inserting it right 'up to the hilt'.

9. Twist the needle gently a few times in both directions for 10-20 seconds.

10. Withdraw the needle smoothly.

11. Dispose of the needle safely.

Possible problems

Problem

A small drop of blood appears when you remove the needle.

Remedy

Wipe it away with a clean tissue and press the site gently for a minute or so. If the site you needled was in your foot keep the leg up until the bleeding has stopped.

Problem

Persistent bleeding (very unlikely).

Remedy

Apply firm pressure and seek medical advice if the bleeding continues.

Problem

Bruise at site of needling.

Remedy

There is probably no need to do anything, but if it is large or painful you can apply ice for about 5 minutes and seek medical advice if necessary.

Problem

A small swelling appears after you remove the needle.

Remedy

Press and flatten it gently with a tissue for a minute or two.

Problem

Needle breakage (extremely unlikely).

Remedy

Try to pull the end out with a pair of clean tweezers or similar imple-ment. If this fails, consult your doctor or a hospital A&E department.

Problem

Infection (extremely unlikely): the acupuncture site becomes painful, red, hot and swollen, and/or red streaks appear running up the limb. Your temperature may be raised.

Remedy

Consult your doctor or a hospital A&E department as soon as possible.

Problem

The needle appears to be 'stuck' (very rare).

Remedy

If the needle seems difficult to remove, wait a few minutes and then it will come out. You may need to pull a little harder than usual. If it still seems to be fixed consult your doctor or a hospital A&E department.

Precautions

- ▷ Use only the needles that have been approved for your use.

- ▷ Check the expiry date on the box and don't use the needles if this date has passed.

- ▷ If any difficulty due to acupuncture occurs please feel free to telephone us to ask for advice.

- ▷ Please follow the instructions given to you exactly, especially as regards frequency of treatment and site(s) of needle insertion. Don't change these without consulting us first.

- ▷ If at any time the treatment ceases to be effective make a fresh appointment.

- ▷ Never put needles into areas of skin that are sore, infected, bruised, or abnormal in any way.

- ▷ If you become pregnant or are trying to become pregnant you should not do acupuncture until you have discussed it with us.

- ▷ If you start taking aspirin, warfarin, or other medicines to 'thin the blood' you should not do acupuncture until you have discussed it with us.

- ▷ Don't try to treat anyone else.

- ▷ If anyone else becomes accidentally injured by one of your needles you should get advice immediately from us or from a hospital A&E department.

Chapter 24

Acupuncture Mechanisms

Chapter Outline

▷ **Current views**

▷ **Recent idea: adenosine**

▷ **Central mechanisms**

▷ **The placebo question**

24.1   Introduction

Since the 1970s there has been a good deal of research to find out how acupuncture works. The early studies mostly focused on the endogenous opioids (endorphins), which had been discovered not long before and which seemed to provide a rational basis for acupuncture analgesia. While these are still part of the story there is much interest these days in other ideas, including the use of the new brain imaging techniques to study central changes in acupuncture.

Although we are still a long way from being able to provide a full explanation for how acupuncture works we can already point to plausible mechanisms for many of the effects, especially pain relief. Adrian White and his colleagues identify five types of mechanism[1].

162

▷ Local changes in the tissues

▷ Segmental analgesia

▷ Extrasegmental analgesia

▷ Central regulatory effects

▷ Myofascial trigger points

As they remark, there are undoubtedly other mechanisms we don't yet know, partly because there is much about the mechanism of pain that we don't know. An example is recent research on the role of adenosine.

24.2 Adenosine

Adenosine is a purine nucleoside which plays an important part in biochemical process such as energy transfer via adenosine diphosphate (ADP) and adenosine triphosphate (ATP). It is an inhibitory neurotransmitter involved in promoting sleep and inhibiting arousal (effects often seen in clinical practice). It also has anti-inflammatory properties. There is now a suggestion that it is involved in the response to acupuncture, at least in mice, which Goldman and colleagues summarise like this.

> We found that adenosine, a neuromodulator with anti-nociceptive properties, was released during acupuncture in mice and that its anti-nociceptive actions required adenosine A_1 receptor expression. Direct injection of an adenosine A_1 receptor agonist replicated the analgesic effect of acupuncture. Inhibition of enzymes involved in adenosine degradation potentiated the acupuncture-elicited increase in adenosine, as well as its anti-nociceptive effect. These observations indicate that adenosine mediates the effects of acupuncture and that interfering with adenosine metabolism may prolong the clinical benefit of acupuncture[2].

24.3 Pain memory

In this context 'memory' refers to a change in an organism that affects its subsequent behaviour. So the immune system provides an example of memory in this sense. When we suffer an infection it often produces changes in the immune system which increase our resistance to that infection subsequently. The immune system 'remembers' the infection.

Not all kinds of memory are as useful to us as is acquired immunity—sometimes it is just the opposite (think of autoimmune diseases). Chronic pain is a form of unwanted memory. This does not depend on the conscious recollection of pain but on long-lasting changes in the structure and function of the central nervous system—brain and spinal cord[3]. So chronic pain—pain that continues long after the initial injury has healed—is due to faulty learning.

When acupuncture relieves the pain in such cases, it does so by causing the pain to be 'forgotten'. Often this requires repeated treatments but it can be almost instantaneous. For example, a man who had had pain in an abdominal scar due to surgery performed over twenty years earlier experienced complete and permanent relief as soon as needles were inserted in the scar.

Pain memory is thought to be due to alterations in synaptic links between neurons, a phenomenon called long-term potentiation (LTP).

24.3.1   LTP

LTP (long-term potentiation) was discovered in 1968 in the rabbit hippocampus. (The hippocampus is a part of the brain that is essential for the formation of short-term memory.) LTP is now known to be widespread in the central nervous system and depends on the strengthening of synaptic links between neurons. Much current theorising about memory is based on LTP.

LTP occurs in the pain pathways, where C-fibre inputs increase LTP in the posterior horn cells[4].

24.3.2   LTD

As well as LTP there is also the contrary phenomenon, LTD (long-term depression), which reverses the effects of LTP. Probably one way in which acupuncture works in chronic pain is by inducing LTD in the pain pathways. But too much stimulation can trigger LTP, which may explain why aggravation occurs.

24.4 Pain and the limbic system

Most of the research on how acupuncture influences the brain has centred on the emotional (affective) component of pain. This is understandable, because the same is true of research on brain processing of pain in general. The limbic and paralimbic areas seem to be particularly important here, especially the anterior cingulate cortex. Acupuncture has been found to reduce activity in these regions, which may account for the clinical observation that acupuncture may reduce the unpleasantness of pain as well as its intensity. Patients sometimes say that they still feel their pain but it troubles them less.

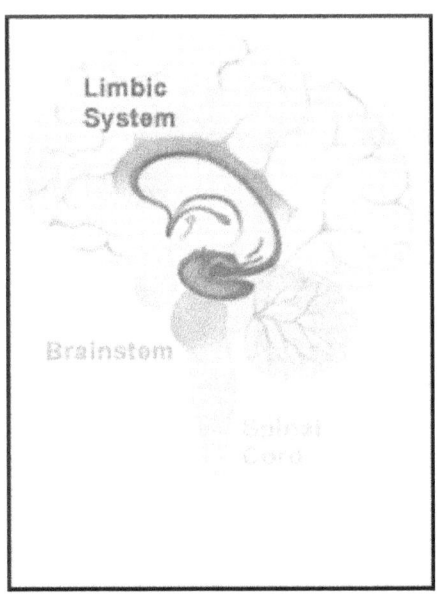

Figure 24.1

24.5 What about the sensory cortex?

The central brain areas are important but they are not the whole story. Recent research on how the sensory cortex processes pain suggests new ways of understanding acupuncture. It has been known for some time that acupuncture stimulates principally the small-diameter myelinated nerve fibres known as Aδ. Fibres of this type from all over the body project to the primary sensory cortex (Figure 24.2) where there is a spatially-organised map of the body. We now know that this map can be modified in various ways, including by touch. This is the basis of *multisensory pain modulation*[5]. The gate theory of pain had already suggested this by proposing that touch could 'close the pain gates' in the posterior horns of the spinal cord and so inhibit the perception of pain[6]. But similar touch-induced analgesia has now been found to occur at all levels of the nervous system.

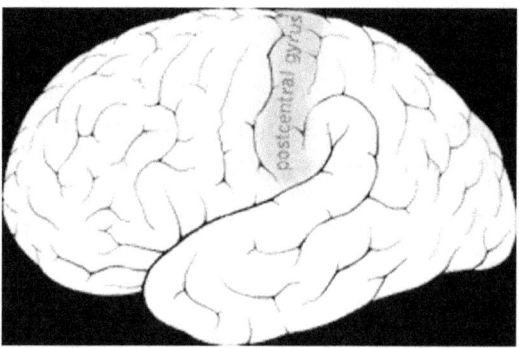

Figure 24.2

24.6   Visual analgesia

Even more remarkable, experiments with visual illusions using mirrors and virtual reality have shown that seeing one's body (or a body image) can reduce the amount of pain that is felt. This has been termed *visual analgesia*[7].

In one study, for example, a mannequin was filmed from behind and projected via a head-mounted display. Subjects saw either this image or one of a neutral object (a large cardboard box). The subjects' backs and that of the mannequin or box were stroked simultaneously with a soft brush. This produced an illusion of body displacement into the mannequin and also raised the subjects' pain threshold, compared with viewing the box.

24.7   Multisensory analgesia

The effects just described constitute *multisensory analgesia*[8]. Acupuncture supplies several different kinds of signal to the brain: touch in the preliminary examination for tender areas; needle stimulation, mainly of Aδ fibres; and sometimes visual input from the patient's sight of the needle insertion. All this suggests that it should be an excellent means of inducing multisensory analgesia. The bigger and more varied the sensory input the greater the pain relief is likely to be. It is possible that acupuncture is more effective if people can see the needles in situ, which would add the element of visual analgesia.

24.8 Just a placebo?

The principal objection raised by critics of acupuncture is that it is simply an elaborate placebo. One answer to this is to point out that placebos must ultimately work by modifying the way the nervous system works[9]. (What else could they do?) This is also what acupuncture does. But we need more than this; we need to be able to point to mechanisms. Thanks to the research I have cited we can now begin to do this. Touch and sight have both been shown to be capable of modifying how the brain processes pain and both are involved in acupuncture.

This means that there is more to acupuncture than just the insertion of needles, but we knew that anyway. Quite apart from whatever we say to the patient by way of scene-setting, there is usually the preliminary examination for TPs, for example, and there may be the sight of the needles being inserted as well. It is very difficult and perhaps impossible to isolate the needle from the rest of the treatment.

So clinical trials using non-penetrating needles or needles inserted superficially at 'incorrect' sites as control procedures become more than ever open to question. But to dismiss acupuncture as working simply by 'suggestion' is to miss out a lot of very interesting and important neurophysiology.

24.9 References

1. White A, Cummings M, Filshie J. An Introduction to Western Medical Acupuncture. Churchill Livingstone (Elsevier). Edinburgh, 2008.

2. Goldman N, Chen M, Fujita T and others. Adenosine A1 receptors mediate anti-nociceptive effects of acupuncture. Nature Neuroscience 2010; 13, 883–888.

3. Sufka K, Chronic pain explained. Brain and Mind, 2000;1(2):155-179

4. Sandkühler J. Understanding LTP in pain pathways. Pain 2007;3:9. Available from http://www.molecular pain.com/content/3/1/9.

5. Haggard P, Janetti DG and others. Spatial sensory organization and body representation in pain perception. Curr Biol 2013;23:164-176.

6. Melzack R, Wall PD. Pain mechanisms: a new theory. Science 1965;150:171-179.

7. Longo MR, Betti V, Agliotti SM and others. Visually induced analgesia: seeing the body reduces pain. J Neurosci 2009;20:12125-12130.

8. Campbell A. Seeing the body: a new mechanism for acupuncture analgesia? Acupunct Med 2013;31:315-318.

9. Campbell A. Hidden assumptions and the placebo effect. Acupunct Med 2009;27:72-75.

Chapter 25

The German Acupuncture Trials

Chapter Outline

▷ **Trial design**

▷ **Trial outcomes**

▷ **References**

25.1   Introduction

These very large-scale trials (among the largest acupuncture trials ever conducted) were set up at the request of the German government in 2001 and published in 2006 and 2007. They were designed to answer the question whether it was justifiable for acupuncture to be paid for by six German statutory health insurance companies.

There were five trials in all: one observation trial to assess the incidence of adverse effects from acupuncture and four randomised controlled clinical trials, for low back pain, knee osteoarthritis, migraine prophylaxis, and chronic tension-type headaches.

25.2   Trial design

The design of all the therapeutic trials was similar. There were three arms: one group received real acupuncture according to TCM principles, another group received sham acupuncture (the needles were inserted superficially at non-acupuncture points), and the third received conventional treatment.

All the physicians who took part were trained in TCM. The patients were blinded to whether they were receiving real or sham acupuncture, and assessment of the results was by blinded judges

25.2.1   Low back pain trial[1]

1267 patients took part in this trial. The results were assessed on standard criteria after six months. Improvement occurred in 47.6% of those receiving real acupuncture, 44.2% of those receiving sham acupuncture, and 27.4% of those treated conventionally. So both real and sham acupuncture were significantly better than conventional treatment but there was no statistically significant difference between real and sham acupuncture.

25.2.2   Knee osteoarthritis trial[2]

1039 patients took part in this trial. Success rates were 53.1% for real acupuncture, 51.0% for sham, and 29.1% for standard treatment. Once again, both forms of acupuncture were significantly better than conventional treatment but there was no statistical difference between real and sham acupuncture.

25.2.3   Migraine prevention trial[3]

960 patients were chosen initially for this trial but 125 withdrew (106 of these were in the standard treatment group). The average reduction in migraine days was 2.3 days in the real acupuncture group, 1.5 in the sham group, and 2.1 days in the standard treatment group. None of these differences was statistically significant.

25.2.4   Tension-type headache trial[4]

Only four patients were willing to have standard therapy (with amitryptiline) so this arm had to be abandoned. For the remaining two arms 409 patients were randomised. An improvement occurred in 33% of the real acupuncture group and 27% of the sham group; the difference was not statistically significant.

25.2.5   Observational study to assess adverse effects

The clinical trials were preceded in 2001-2005 by an observational study involving 12,617 physicians. Roughly 2.6 million(!) patients were treated in this period. A random sample of 190,924 was selected for review. Adverse events were reported in 7.5% of cases, of which 45 were judged to be serious. The three most frequent effects were bruising at the site of needle insertion, temporary worsening of the existing symptoms, and vasovagal responses.

25.3   Outcome of trials

As a result of this work the German government decided to approve acupuncture for insurance purposes for low back pain and knee osteoarthritis

25.4   References

1. Haake M, Müller, H-H and others. German Acupuncture Trials (GERAC) for chronic low back pain: randomized, multicenter, blinded, parallel-group trial with 3 groups. Archives of Internal Medicine 2007 167 (17): 1892–1898.

2. Scharf H-P, Mansmann U and others. Acupuncture and knee osteoarthritis: a three-armed randomized trial. Annals of Internal Medicine 2006 145 (1): 12–20.

3. Diener H-C, Kronfeld K and others. Efficacy of acupuncture for the prophylaxis of migraine: a multicentre randomised controlled clinical trial. The Lancet Neurology 2006;5(4): 310–316.

4. Endres HG, Böwing G and others. Acupuncture for tension-type headache: a multicentre, sham-controlled, patient-and observer-blinded, randomised trial. The Journal of Headache and Pain 2007 8(5): 306–314.

Chapter 26

Books, Papers and a Journal

Chapter Outline

▷ **Further reading**

▷ **Chapter notes**

▷ **The BMAS and its journal**

26.1 Introduction

Acupuncture is a practical business and the best way to become good at it is to use it a lot. But it is likely that you will want to deepen your knowledge by reading. Most books on acupuncture describe the traditional system but here I mention some that arc relevant to modern acupuncture and its differences from TCM. As for journals, nearly all are traditionalist in character apart from *Acupuncture in Medicine,* which is the journal of the BMAS and which I also describe briefly here.

26.2   FURTHER READING

26.2.1   Felix Mann (1.9)

Felix Mann's revolutionary view of acupuncture is described in his book *Reinventing Acupuncture: A New Concept of Ancient Medicine* (2nd edition 2000; Butterworth-Heinemann, Oxford). It covers periosteal acupuncture, strong reactors, 'dosification', and repetition of treatment. It also describes Mann's approach to treating a wide range of disorders.

This book is, I should say, essential reading for anyone who wants to understand the full extent of the new view of acupuncture that Mann introduced. It is very readable and contains a remarkable amount of first-hand observation of the effects of acupuncture in different conditions.

26.2.2   Bridie Andrews (1.13)

Bridie Andrews has shown that the prevailing Western idea of acupuncture as a traditional treatment that has existed for at least 2000 years needs to be pretty radically revised. By the nineteenth century it had ceased to be practised by educated Chinese physicians and its use was confined to illiterate healers. It was revived in the 1930s by Dr Cheng Dan'an, who was strongly influenced by Japanese ideas. The use of fine needles came from Japan and so did the basing of acupuncture on modern (Western) anatomy. Andrews's book *The Making of Modern Chinese Medicine 1850-1960* (2015; University of Hawai'i Press) has radically altered my understanding of how acupuncture has developed in the modern era. She shows that acupuncture, far from being an ancient art that has persisted from time immemorial, dates only from the 1930s, at least in the form that we recognise today. (For more details see Campbell A, Acupunct Med 2015;33:491-495 doi:10.1136/acupmed-2015-010991.)

26.2.3   Shigehisa Kuriyama

Have you ever wondered why the depictions of acupuncture channels in the older Chinese texts show figures who look so plump?

Figure 25.1

This must surely reflect an important difference in outlook between East and West, but perhaps not exactly what most people tend to assume.

Shigehisa Kuriyama gives the answer in his very interesting book, *The Expressiveness of the Body and the Divergence of Greek and Chinese Medicine* (Zone Books, New York, 1999). In part the reason for the difference is that the Chinese did not recognise the existence of muscles until the 20th century under the influence of Western medical missionaries.

This is less surprising than may at first appear, for the muscles are not really all that obvious. Even in the West it takes exposure to art before we think of the body in this way. The perception of the body as, ideally, a system of well-defined musculature is learned not innate. Go to the beach or the swimming pool and you will see few bodies that demonstrate the 'ideal' delineation of the musculature that is supposed to exist. In most of us the muscles are concealed by subcutaneous fat.

There was no real interest in anatomy in ancient China. Dissection was hardly practised, and even when it was it was not done for scientific but for metaphysical reasons. The same is true of other ancient societies, such as India and Egypt. What is surprising is not that anatomy was not studied at those times but that it did arise in the Enlightenment in Europe.

26.2.3.1 Plumpness and health in ancient China

Another reason why the traditional figures in TCM books are plump is that it was thought to be a sign of health. The earliest theory of acupuncture in China was based on the idea that disease came about because certain kinds of wind ('empty' winds) were harmful and could enter the body via pores. Acupuncture was supposed to work by opening the pores and letting the wind out. The plump figure is healthy because by being stout he can resist the entry of the harmful wind.

Kuriyama brings out a great number of Western misconceptions about TCM which I have not found discussed anywhere else.

26.2.4 Nigel Wiseman

Westerners' ideas about TCM are often wrong. They impute values to it that it doesn't have. For a good detailed discussion of these misconceptions see Nigel Wiseman's paper *Westerners' Alternative Health-Care Values Eclipsing a Wealth of Knowledge* (http://www.paradigm-pubs.com/sites/www.paradigm-pubs.com/files/files/AltHeal.pdf).

> I question certain widely held views about Chinese medicine, namely that Chinese medicine is natural, holistic, caring, nonmechanistic, and even spiritual in nature. In discussions of Chinese medicine outside of Asia, these qualities are often tacitly assumed as facts. In this paper, I trace many of these assumptions to influences of the desiderata of alternative health-care, which tend to define Chinese medicine as being what Western medicine is perceived not to be.

26.2.5 Travell and Simons (5.12)

Although these authors don't write about acupuncture it is impossible to talk about the modern version of the treatment without mentioning their large two-volume work.

▷ Simons DG, Travell JG, Simons LS. *Myofascial Pain and Dysfunction: The Trigger Point Manual*. Vol. 1. Upper Half of the Body. Second Edition. Lippincott Williams & Wilkins, Philadelphia, 1999.

▷ Travell JG, Simons DG. *Myofascial Pain and Dysfunction: The Trigger Point Manual*. Vol.2. The Lower Extremities. Lippincott Williams & Wilkins. Philadelphia.

The anatomical illustrations in these books are outstandingly good and are useful for acupuncturists on those grounds alone.

26.2.6 Peter Baldry

P.E. Baldry's *Acupuncture, Trigger Points, and Musculoskeletal Pain* (3rd edition; Elsevier, Edinburgh, 2005) is, based on the view that modern acupuncture is almost wholly the treatment of TPs. The book covers both research and practice.

26.2.7 White, Cummings, Filshie

Adrian White, Mike Cummings, and Jacqueline Filshie are all long-standing members of the BMAS with wide experience of the subject. Their book, *An Introduction to Western Medical Acupuncture* (Churchill Livingstone, Edinburgh, 2008), is mainly practical but includes chapters on acupuncture mechanisms and on TCM. They make use of acupuncture points although without the requirement for much precision in locating them. They write with a refreshing lack of dogmatism.

26.3 CHAPTER NOTES

26.3.1 Trigger points reconsidered (5.4.1)

▷ Tough EA, White AR and others. Variability of criteria used to diagnose myofascial trigger point pain syndrome—evidence from a review of the literature. Clinical Journal of Pain 2007 23(3);278-276.

▷ Tough EA, White AR and others. Acupuncture and dry needling in the management of trigger point pain: A systematic review and meta-analysis of randomised controlled trials. European Journal of Pain 2009;13:3-10.

26.3.2 Acupuncture in pregnancy (6.9)

▷ Wedenberg K, Moen B, Norling A. A prospective randomized study comparing acupuncture with physiotherapy for low back pain and pelvic pain in pregnancy. Acta Obstetrica et Gynaecologica Scandinavica 2000;79(5):331-335.

▷ Elden H, Lasfors K, Olsen MF and others. Effects of acupuncture and stabilising exercises as adjunct to standard treatment in pregnant women with pelvic girdle pain: a randomised controlled trial. BMJ 2005;220(7494):761-766.

▷ The safety of acupuncture in pregnancy: a systematic review. Park J, Youngjoo S, White AR. Acup med 2014;21:257266.

26.3.3   Safety review (6.12)

▷ White AR, Hayhoe S and others. Adverse events following acupuncture: prospective survey of 32 000 consultations with doctors and physiotherapists. BMJ 2001;323:485. Also, see 25.2.5.

26.3.4   *De qi* (7.2.2)

▷ White P, Prescott P, Lewith G. Does needling sensation (*de q*i) affect treatment outcome in pain? Analysis of data from a larger, single-blind randomised controlled trial. Acupunct Med 2010;28:120-125.

26.3.5   Shoulder corticosteroid injection (14.1)

▷ Buchbinder R, Green S, Youd JM. Corticosteroid injections for shoulder pain. (onlinelibrary.wiley.com, doi:10.1002/14651858.CD004016/full.)

26.3.6   Epicondylitis (15.2)

▷ Rothschild B. Mechanical solution for a mechanical problem: Tennis elbow. World J Orthop 2013;18(4):103-106.

26.3.7   Acupuncture for knee pain (19.1)

▷ White A, Foster NE Cummings M and others. Acupuncture treatment for chronic knee pain: a systematic review. Rheumatology (Oxford (2007);46(3):384-390.

26.3.8 Phantom limb pain (20.3.7)

▷ Bradbrook D. Acupuncture for phantom limb pain and phantom sensation in amputees. Acupunct Med 2004;22:93-97.

▷ Davies A. Acupuncture treatment of phantom limb pain and phantom limb sensation in a primary care setting. Acupunct Med 2013;31:101-194.

26.4 BMAS JOURNAL

Acupuncture in Medicine is the journal of the BMAS. It is published six times a year by the BMJ Publishing Group. This gives it a degree of academic respectability which is unmatched by any other acupuncture journal. It is wholly concerned with the modern understanding of acupuncture and is essential reading for anyone who has a serious interest in modern medical acupuncture.

26.5 The BMAS

The BMAS was founded in 1980 as the successor to Felix Mann's informal group made up of his former students; he was its first President. Membership is open to all health professionals who are regulated by statute in the United Kingdom. It is not necessary to have done a training course to become a member.

The BMAS provides training of various kinds and holds scientific meetings twice a year at which papers on acupuncture and related topics are presented. For details see their website (www.medical-acupuncture.co.uk).

26.5.1 BMAS Certification

The BMAS offers two grades of certification to its members, the Certificate of Basic Competence (CoBC) and the Diploma of Medical Acupuncture (DipMedAc). They are not based on examination but on demonstration of awareness of the risks of acupuncture and the submission for approval of a certain number of case studies. While these certificates indicate a certain level of training and experience they are not statutorily recognised acupuncture qualifications, which do not exist at present.

Acupuncture is not regulated in Britain at present so anyone is free to call themselves an acupuncturist, with or without any form of training.

Index

www.ingramcontent.com/pod-product-compliance
Lightning Source LLC
Chambersburg PA
CBHW080911170526
45158CB00008B/2070